PHILOSOPHIÆ NATURALIS

PRINCIPIA MATHEMATICA

Revision IV

By

Isaac Newton

And

Keith Dixon-Roche

PHILOSOPHIÆ NATURALIS PRINCIPIA MATHEMATICA Revision IV

Keith Dixon-Roche © 2017 to 2019

All concepts and formulas in this book not previously

attributed to 'The Heroes' identified in Appendix A8,

are the sole property of Keith Dixon-Roche

and protected by copyright.

Their use, publication, broadcasting,

distribution, copying or any form of recording

without Keith Dixon-Roche's written consent

shall be a breach of international copyright law

and subject to legal action.

PHILOSOPHIÆ NATURALIS

PRINCIPIA MATHEMATICA

Revision IV

Published by CalQlata
info@CalQlata.com

First published November 2018
Second publication February 2019
Third publication June 2019
Fourth publication December 2019
Copyright © Keith Dixon-Roche 2018

This book is sold subject to condition
that it shall not by way of trade or otherwise,
be lent, re-sold, hired out or otherwise circulated
without the publisher's prior consent and in such
circumstances it shall not be circulated in any form of
binding or cover other than that in which it is published

PHILOSOPHIÆ NATURALIS PRINCIPIA MATHEMATICA Revision IV

Contents

Preface			**11**
1	**Introduction**		**13**
	1.1	What Went Wrong	14
	1.1.1	The Photon?	16
	1.2	And Now?	17
	1.3	Where Do We Go From Here?	19
	1.4	How This Book Is Organised	21
2	**Narrative**		**23**
	2.1	Energy	25
	2.1.1	Electricity	28
	2.1.2	Magnetism	31
	2.1.3	Electro-Magnetic Energy	34
	2.1.4	Potential	39
	2.1.5	Kinetic	40
	2.1.6	Heat & Temperature	41
	2.2	Orbits	43
	2.2.1	The Laws	49
	2.2.2	Elliptical	50
	2.2.3	Circular	52
	2.2.4	Force-Centre Mass	53
	2.2.5	Planetary Mass	54
	2.2.6	External Influences	56
	2.2.7	Centrifugal Force	57
	2.2.8	Station-Keeping	58
	2.2.9	Orbital Planes	59
	2.2.10	The Importance of Orbits	60
	2.3	Spin	61
	2.3.1	Polar Moment of Inertia	64
	2.3.2	Earth's Core	65
	2.3.3	Earth's Magnetic Field	66
	2.3.4	Magnetic Reversal	67
	2.3.5	No Moon!	68
	2.3.6	Goodricke & Algol	69
	2.3.7	Chicken & Egg	70

2.4	Core-Pressure		71
	2.4.1	The Structure of Celestial Bodies	72
2.5	The Atom		73
	2.5.1	Atomic Particles	74
	2.5.2	Proton-Electron Pair	78
	2.5.3	Electron Shells	80
	2.5.4	Nucleus	81
	2.5.5	Electron & Proton Spin	82
	2.5.6	Isotope	83
	2.5.7	Ion	84
	2.5.8	Radioactivity	85
	2.5.9	How They Work	87
	2.5.10	The State of Matter	90
	2.5.11	Fission & Fusion	91
	2.5.12	The Early Atom	93
2.6	The Universe		95
	2.6.1	Life, the universe & everything	97
	2.6.2	The Milky Way	99
	2.6.3	Our Solar System	101
2.7	Fact & Fiction		117
	2.7.1	Sub-Atomic Particles	118
	2.7.2	Black Holes	119
	2.7.3	Big-Bang	121
	2.7.4	Dark Matter	123
	2.7.5	The Birth of Our Solar System	124

3 Calculation Procedures — 127

3.1	Energy		129
	3.1.1	Electrical	130
	3.1.2	Magnetic	131
	3.1.3	Electro-Magnetic	132
	3.1.4	Potential	133
	3.1.5	Kinetic	134
3.2	Orbits		135
	3.2.1	Laws	136
	3.2.2	Elliptical Orbits	137
	3.2.3	Circular	145
	3.2.4	Force-Centre Mass	157

		3.2.5	Planetary Mass	158
		3.2.6	External Influences	159
		3.2.7	Centrifugal Force	160
		3.2.8	Station-Keeping	161
	3.3	Spin		163
		3.3.1	Polar Moment of Inertia	167
		3.3.2	Earth's Core	168
		3.3.3	Earth's Magnetic Field	169
		3.3.4	Magnetic Reversal	170
		3.3.5	No Moon!	171
		3.3.6	Hades	172
	3.4	Core-Pressure		175
	3.5	The Atom		179
		3.5.1	Atomic Particles	181
		3.5.2	Electron Shells	186
		3.5.3	Nucleus	187
		3.5.4	How They Work	189
4	**Calculation Results**			**193**
	4.1	Energy		195
		4.1.1	Electricity	196
		4.1.2	Magnetism	197
		4.1.3	Electro-Magnetic Energy	198
	4.2	Orbits		199
		4.2.1	Galactic	200
		4.2.2	Solar	202
		4.2.3	Lunar	204
		4.2.4	Atomic	206
	4.3	Spin		209
		4.3.1	Polar Moment of Inertia	210
		4.3.2	The Earth's Core	211
		4.3.3	The Earth's Magnetic Field	212
		4.3.4	Our Sun	213
		4.3.5	Our Planets	214
		4.3.6	Hades	215
	4.4	Core Pressure		217
	4.5	The atom		219

	4.5.1	Cold (T_c)	220
	4.5.2	Planck Minimum (T_o)	221
	4.5.3	Planck Mean (T_m)	222
	4.5.4	Neutronic (T_n)	223

5 The Physical Constants — **225**

5.1	Introduction		227
5.2	Symbols		229
5.3	Physical Constants		233
5.4	General Physical Constants		235
5.5	Universal Heat & Charge Capacities		237
	5.5.1	Microstates	238
5.6	Specific Heat Capacities (particles)		239
5.7	Specific Charge Capacities (particles)		241
5.8	Electricity		243

6 Support — **245**

6.1	Proof of the Orbital Model		247
	6.1.1	Nicolaus Copernicus	247
	6.1.2	Johannes Kepler	247
	6.1.3	Galilei Galileo	247
	6.1.4	Isaac Newton	250
6.2	The Problem with Relativity		267
	6.2.1	Light Deflection	268
	6.2.2	The Speed of Light	271
	6.2.3	Neutronic Radius (R_n)	272
	6.2.4	Elliptical Orbits	273
	6.2.5	$E=mc^2$	275
	6.2.6	Hades	278
6.3	The Problem with Quantum Theory		279
6.4	The Error		281
	6.4.1	Measured Vacuum	282
6.5	Model Verification		283
	6.5.1	Density vs Temperature	284
	6.5.2	Specific Heat Capacity	285
	6.5.3	Gas-Point	286

	6.5.4	Our Sun	287
	6.5.5	PVRT	288
6.6	Heat		289
6.7	Mass		291
6.8	Gravity		293
6.9	Electricity		299
6.10	Newton vs Planck		303
	6.10.1	Newton's Atom	305
	6.10.2	Planck's Atom	306
6.11	Important Constants (explained)		307
	6.11.1	Σ	308
	6.11.2	G	319
	6.11.3	φ	312
	6.11.4	k, k', μ_o, ε_o	313
	6.11.5	h, h'	314
	6.11.6	e, e'	317
	6.11.7	R_∞, R_γ, a_o	318
	6.11.8	ρ_u	319
	6.11.9	R_s	320
	6.11.10	R_n	321
	6.11.11	RAC & RAM	322
	6.11.12	N_A	323
	6.11.13	k_B, k_B'	324
	6.11.14	K	325
	6.11.15	v_C & v_E	326
6.12	The Laws of Thermodynamics		327
7	**Things You Can Do**		**329**

Appendices **331**

    A1 General   333

    A2 References   335

    A3 Glossary   337

    A4 Symbols   339

    A5 Useful Formulas   341

    A6 Corollaries   343

    A7 Hypotheses   347

    A8 The Heroes   349

Preface

I have always believed that if a mathematical law applies to one feature of nature it must apply to all of it: i.e. a law must by definition, be universal. I also feel that science took a wrong turning in the first quarter of the twentieth century owing to the dissemination of highly speculative theories that were accepted simply because of the prominence of their proposers. However, I was not sufficiently familiar with the subject to dispute it. After two and a half years of detailed study, that situation has changed and it appears to me that a hundred years may have been wasted in the search for impossible solutions. Isaac Newton's laws should have prevailed.

Newton apparently devised his theories to settle a bet, and like everything he tackled he took this work seriously. Despite having only Kepler's elliptical orbits and Galileo's laws of motion at his disposal, Newton managed to develop an all-encompassing theory that remains universally valid today. It was published in three revisions between 1687 and posthumously. He published only because of the persistence of one of his few friends: Edmund Halley. It is for this, rather than for his comet, that we owe Edmund Halley our deepest gratitude.

Newton's gravitational theory is complete and totally accurate. It covers all the bases. His model relies on a concept he called his 'constant of motion' to keep things moving. However, even he didn't realise that his theory also applies to circular orbits in which a satellite (e.g. an electron) provides its own kinetic energy, or that his gravitational constant (G) may be used to calculate the deflection of electro-magnetic radiation (light).

His laws of gravitation and motion together describe the behaviour of everything in the universe from atomic particles to the Big Bang, and they do so with absolute simplicity and accuracy, except for one small omission; he did not explain spin theory, without which it is difficult to explain all motion. However, it would have been very difficult for him to have developed this theory with the limited information and facilities available to him.

A couple of years ago, my daughter gave me a copy of Colin Pask's *Magnificent Principia* (Pask; 2013). After reading it, I was left with the suspicion that there were many unanswered questions about Newton's discoveries and I wondered how much had been done to continue Newton's work during the subsequent 300 years. Very little it seems.

I therefore set out on my quest to prove every aspect of Newton's theory of orbital motion, and see if I could determine the source of planetary spin. Having completed these objectives, I continued with core pressure, the earth's magnetic field, the definition of his gravitational constant 'G' and finally the atom, all using his theories.

After completing my model of the atom and having discovered how it really works, I was stunned by its simplicity and brilliance. Its existence must surely be due to providence, not chance. If there is one thing that could prove the existence of a being of supreme intelligence, and I am not referring to anybody's particular god; it is the atom. In the immortal words of a great contemporary philosopher; I was that *"girl sitting on her own in a small café in Rickmansworth"* (Adams; 1980), and I couldn't understand why none of this had been done before.

I am an engineer, not a scientist. Whilst I have always had an interest in science, I never had the opportunity to study it in detail. As a non-scientist, I have been able to tackle the subject free from the dogma that the scientific community has acquired since it displaced the religious community's hegemony over its own flawed natural laws.

Whilst my theories and models may not be perfect, everything in them can be supported with known scientific theories evolved well before the twentieth century.

I realise of course, that just as with Copernicus, Kepler, Galileo, Newton and Wegener before, none of these findings will be appreciated whilst the current scientific community exists. That august body is hardly likely to accept theories that disprove those for which they have been awarding themselves so many prestigious prizes. My hope is that maybe, one day, a new generation of free-thinking scientists will discover this work, correct, complete and advance it, and in so doing get science back on track.

Because it is now possible to define the Milky-Way's force-centre, I have given it the name 'Hades' for easier reference.

Keith Dixon-Roche 2018

PHILOSOPHIÆ NATURALIS PRINCIPIA MATHEMATICA Revision IV

1 Introduction

Science got itself into a bit of a mess during the 20th century owing to a couple of obscure theories, neither of which can be reconciled with concepts that we *know* work, but which stubbornly refuse to go away. Together these theories have inspired countless myths that simply multiply with the passing years. Nobody appears to be questioning them and nobody is able to verify them.

It has now become standard practice within the scientific community to justify any irreconcilable theory simply by claiming that *"the laws of physics do not apply"*.

So, I decided to have a go myself, by starting all over again; going back to basics (the year 1900).

Apart from Max Planck's assistance, I have managed to sort out this mess and compile a complete working theory for the universe using principles that were available well before 1900.

I have also managed to describe all the universal constants (including electrical) in the same basic units of energy (mass, length, time, charge & temperature; refer to Chapter 5)

Anything in this book that has not been fully resolved (and there isn't much) is referred to as *hypothesis*.

Whilst my hypotheses are perfectly robust, they remain as such because a couple of details need confirming/correcting. The contentious aspects mostly involve the nature of neutrons, but this has not been addressed here because it has nothing to do with Newton's laws of orbital motion.

Unresolved issues are highlighted in the text with the superscript [?] in which '?' will be replaced with a number that can be found in Chapter 7

1.1 What Went Wrong

Unfortunately, about a hundred years ago, a prominent scientist stated of his own theoretical model: *"if you aren't profoundly shocked by quantum physics, then you haven't understood it"*
Another did not appear to understand the basis for Henri Poincaré's formula E=mc² and actually declared to Georges Lemaître that his (Lemaître's) *"science was not very good"*

Such comments should be treated with extreme caution ...
... **if there is one thing certain about nature, it does not need to rely on complexity for an elegant solution**, and scientific laws *never* rely on statistics because statistics are subject to change; *laws are not.*
Statistics are akin to chaos theory: they are a means of guesswork used in situations where insufficient information is available to explain events accurately. They apply to the consequences of laws, not the laws themselves.

Quantum theory is inelegant, over complicated, reliant on statistics, cannot be reconciled with Newton's laws of orbital motion, cannot emit energy and remains unresolved after a hundred years. It is highly likely therefore, that it is nothing more than an obscure theoretical exercise.

Relativism can be disproved using Newton's gravitational constant and dark matter remains undiscovered. Poincaré's formula has nothing to do with kinetics. Classical atomic theory appears to be incorrect. Black-holes are wrongly said to be singularities that spin at the speed of light. Nobody has tried to determine the source of planetary (and therefore atomic) spin or core-pressure.

Isaac Newton pointed us in the right direction 300 years ago, but since the early 20th century the entire scientific community seems to have discounted the suitability of his theories for the evaluation of atoms (quantum theory) and galaxies (dark matter) simply because a couple of well-known scientists took this view. For example:

Relativism appears to have been partly based upon the supposition that $E=mc^2$ applies to kinematics, whereas it is a limiting case for potential energy based upon Newton's and Coulomb's laws and the creation of neutrons. Moreover, it incorrectly assumes that light possesses mass.

It is incorrectly currently believed that mass converts to energy with speed.

PHILOSOPHIÆ NATURALIS PRINCIPIA MATHEMATICA Revision IV

Dark matter in the form of sub-atomic particles was postulated because Newton's laws were said to predict a great deal more matter in the Milky Way than appears through observation. This has been easily disproved.

It was long ago assumed that we need sub-atomic particles (e.g. quarks, leptons, fermions, bosons, gluons, etc.) to hold atomic particles together and make the atom work. It now appears that none of these are necessary.

We have been taught that atomic shells are elliptically flat, can hold more than two electrons, and that each electron within a shell is in some way different from all others. It now appears that this level of complexity is unnecessary.

As Newton's gravitational constant (G) is based upon Quanta why shouldn't his theories also apply to atoms?

We have been advised by the world's most eminent astrophysicists that it is impossible to calculate spin in satellites and force-centres. Yet Newton's laws provide us with all the information needed to solve this problem.

Nobody appears to have grasped the fact that Newton's formula directly (with no reinterpretation) allows us to calculate the pressure inside a solid body, such as a planet or star, so why are we still guessing internal pressures?

In fact, guesswork appears to be prevalent throughout science today.

Together with the help of a number of early heroes (refer to Appendix A8), Newton provided everything we need to understand our universe ...
... how it was created,
... the age of everything in it,
... how it works,
... what everything in it is made of,
... how it generates its energy and
... where it stores this energy;

The end of the 18th century saw the start of the industrial revolution, which continues today, only now; it is called a technological revolution. The start of the 20th century should have kicked off a scientific revolution. It never happened. Why?

1.1.1 The Photon

The problem was the photon.

I need to deal with this issue now in order that it doesn't interfere with your understanding of the universal model discussed in this book.

It is about time we all dropped the concept of photons, i.e. the belief that electrons travelling at the speed of light emit light; *they don't*.
We have been taught this for a hundred years, forcing us to create weird and wonderful theories to explain how *mass* moves in waves; *it doesn't*.
The photon exists in our minds because of a very simple mistake made a long time ago related to Crooke's tube (refer to Chapter 6.4).

Once this is understood, the whole problem of energy, magnetism, gravity, electricity, etc. vanishes. You can ignore quantum theory and the theory of relativity, both of which were invented to explain this misunderstood behaviour of electrons.

The deflection of light can *only* be explained using Newton's gravitational constant (G), and the behaviour of electrons within atoms can *only* be resolved using Newton's laws of orbital motion and Coulomb's laws of electrical force (refer to Chapters 6.2.1 & 6.11.2). We should not, however, forget William Gilbert's contribution, which predates and forms the basis of all the theories related to force and energy fields (both atomic and astronomic).

It appears to me that if everybody had realised that Crooke could not possibly have created a perfect vacuum in his tube, we would not have been confused by quantum theory and the theories of relativity, and we would now be 100 years into a *'scientific revolution'*.

1.2 And Now?

Whilst the theories proposed in this book concerning Newton's Laws of Orbital Motion, Orbital Systems, Planetary Spin, Core Pressure, the Atom and Earth's Magnetic Field are a matter of scientific fact, those on Energy and the universe are hypotheses.

However, they ...
... are based on and obey well-known universal laws of nature that work
... have no need for statistics, unification theories or obscure concepts
... reflect what we sense in the universe
... have no need for intimidation

It cannot have escaped everyone's notice that Newton's, Coulomb's, Gilbert's, Maxwell's and others' force formulas all have the same configuration:
$F = K.v_1.v_2 / R^2$ (which is actually: $F = K.v_1.v_2 / A$)
where: 'K' is a constant, 'v' a variable and 'A' the spherical surface area at radius (R).

My own calculations have revealed a similar relationship for the conversion of electro-magnetic energy to velocity in electrons:
$T = X.v^2 / e^2$
where 'X' is a constant, 'v' the velocity of an electron and 'e' its electrical charge.

If all these formulas *look* the same, they probably *are* the same, i.e. they are simply variations based upon our current misunderstanding of gravity, mass, heat, etc. which are actually the same thing; *energy*. Thus, there are really only two formulas, one of which is for electrical force (Coulomb) and the other for magnetic force (Gilbert/Newton) that differ by a coupling ratio (φ = 4.407E-40). Given that gravity is magnetism, we need be in no doubt that Newton's formula represents magnetic force and can be explained as such (refer to Chapter 6.8).

The atomic model proposed here is elegant, eternal, predictable and brilliantly simple; anyone can understand it without the need for shock-tactics. It also complies with all of Newton's, Gilbert's, Coulomb's, Faraday's and Maxwell's laws, so there is no need for unification theories or statistics. In fact, it now looks highly likely that contrary to popular scientific opinion, these laws are sufficient to explain everything in our universe. Newton's laws are indeed universal, and via them, we can create realistic solutions for virtually everything in our universe from atomic to astronomic physics, including: neutronic energy, '*Big-Bang*', Earth's magnetic field, 'G', 'E=mc²', ultimate density and a great deal more.

Everything is energy: our universe is very much simpler than the one we have been taught, and exploited properly it can provide us with all the clean, free energy we need, simply from Newton's orbits.

If my model is correct (or even close), it then becomes a simple, albeit time-consuming enterprise to determine everything there is to know about our universe, from the very smallest (Quanta) to the very largest (*Big-Bang*) using theories that have been known since Poincaré first revealed his formula and Crooke discovered electro-magnetic energy in the 19th century.

1.3 Where Do We Go From Here?

Given what we now know about universal energy;
1) How it is created (orbits and spin-friction)
2) Where it is created (stars and planets)
3) How it is transmitted (electro-magnetic energy)
4) Where it is stored (neutrons)

We now have access to unlimited, clean, free energy sources;
1) Elliptical Orbits
2) Mantle heat
3) Neutrons

Moreover, these theories can give us the ability to *mathematically* predict chemical reactions in *all* matter irrespective of complexity; the *ultimate calculator*.

Such a calculator would preclude the need for material, chemical or pharmaceutical testing and experimentation. No more risk, material, time or money need be wasted on such activities and every country in the world would be able to design [100% accurate] new materials, chemicals and medicines in safety, from a computer terminal with trained but semi-skilled personnel. Furthermore, the creation of comprehensive organic and inorganic chemical databases will remove the need for duplicate effort together with the horrendous qualification periods for new medicines imposed by various national and international health authorities.

Because we now know where the universe stores its energy, we have access to an unlimited supply free from waste and pollution. We could do something useful with the world's nuclear waste; as the fuel for clean, controllable, efficient energy generators of any size. Much less mining!

Moreover, due to the discovery of the true meaning of $E=mc^2$ (refer to Chapter 6.2.5), there is no longer any reason to assume that light-speed is a limiting condition for matter. And if matter has no mass, imposing a limiting velocity owing to the conversion of mass to energy becomes unnecessary. The speed of light is simply a speed for electro-magnetic radiation, such as that for sound: there's no reason it cannot be exceeded.

Anti-gravity also becomes *theoretically* possible. All you need to do is repel the earth's *magnetism*, which is easier than opposing *gravity* with mass.

PHILOSOPHIÆ NATURALIS PRINCIPIA MATHEMATICA Revision IV

A few of the possibilities from the discoveries explained in this book are listed below?

1) Molecular calculator (and database) giving new (perfect) materials, medicines and chemicals in minutes
2) Clean, free efficient energy (by-product = hydrogen)
3) Propulsion-free satellites
4) The ability to safely recycle nuclear waste
5) Energy cells that can be fuelled with any matter (e.g. rocks!)
6) Alter elements into something else
7) Change the colour of matter electrically
8) Together with PERS#, the elimination of skin-friction offers virtually free travel
9) Perfect lubricants (machines with almost eternal life)
10) Free energy from the earth's mantle
11) Massive reductions in: pollution, material waste, energy, etc.
PERS = *potential energy recovery system*

In other words, we now have the ability to ...
... massively reduce energy and battery production;
... massively reduce mining requirements;
... massively reduce transport costs;
... massively reduce the number of chemical laboratories;
... eliminate; national power stations & transmission lines, wind-turbines & solar panels;
... eliminate pollution from energy generation;
... create vehicles with no engine or drivetrain that need no refuelling;
... create 100% recyclable packaging

All the energy we use today requires the generation of much more to harness and recycle it. Instead of generating energy at an efficiency of less than 10%, we now have access to energy generation that is 231,000,000% efficient.

Instead of swapping one pollutant for another and/or simply moving it around as we do today, we could now create a genuinely clean place for everyone in which to live; together with limitless cheap energy for all.

1.4 How This Book Is Organised

This book comprises 8 sections, the first four of which provide similar information but in a different form:

2 Narrative

A written description that gives a general overview of the various discoveries made in this book. It is devoid of formulas and mathematical complexity with a view to providing a *'light-read'*!

3 Calculation Procedures

A compilation of the mathematical formulas supporting the narrative, including how to use them. This section has been written to simplify their use.

4 Calculation Results

A collection of [mostly] tabulated calculation results for selected examples using the formulas provided in section 3 (above).

5 Physical Constants

All the physical constants (including electrical properties such as Volts, Amps, Henries, Farads, Ohms, etc.) are provided (to ≤15 decimal places) in terms of the same four basic units; length, time, mass and charge and two ratios: m_e, e, R_n, t_n & ξ_v, ξ_m

6 Support

A mathematical and descriptive explanation for all the physical constants and scientific discoveries along with the reasons why Relativity and Quantum Theory must now be discarded.

7 Things You Can Do!

A list of unresolved issues.

8 Appendices

References, symbols, glossary, etc. used throughout this book along with a summary list of corollaries and hypotheses.

PHILOSOPHIÆ NATURALIS PRINCIPIA MATHEMATICA Revision IV

2 Narrative

A written description that gives a general overview of the various discoveries made in this book. It is devoid of formulas and mathematical complexity with a view to providing a *'light-read'*!

PHILOSOPHIÆ NATURALIS PRINCIPIA MATHEMATICA Revision IV

2.1 Energy (hypothesis)

What is energy and how does it apply to the universe we know today?

First; we need to understand that *everything* in the universe is composed of electrical and magnetic energy and nothing else; there is no such thing as gravity, mass, heat, etc.

Energy cannot be lost or gained and it can only be transferred by electro-magnetic radiation.

Energy packets are what we understand as atomic particles and what Max Planck referred to as Quanta, which is the term that will be used in this book to collectively describe the only two that are necessary to make the universe work; the proton and the electron. Neutrons are protons and electrons combined through high temperature.

The elimination of **mass** from science is difficult to accept along with everything else, so whilst I explain the concept (refer to Chapter 6.7), I shall continue to refer to it throughout this book as '*mass*' to minimise confusion for the reader.

Energy was not a concept known to Isaac Newton so he used force to describe energy transfer.
Force is the manifestation of 'energy transferred between two or more bodies separated by physical distance.
This relationship is better known as; Energy = Force x Distance.
The difference between the two concepts can be described thus:
Energy is a force applied over a distance
Force is energy per unit distance

Apart from the non-polar magnetic charge present in all atomic particles, magnetism and electricity are polar; negative or positive. Opposite poles attract and similar poles repel.

Both types of energy (electrical and magnetic) have static and dynamic counterparts. Because Quanta have opposite electrical and magnetic polarity, when encountering other Quanta, their electro-magnetic energies will *always* be opposite *or* identical according to nature's requirements; i.e. polarity conflict is impossible.

How does this relate to what we see and feel?

Low-temperature scenario: when you see an object, such as a cup, you are seeing all the adjacent atoms in that cup held together with magnetic field energy. In this form, the atoms are sufficiently close together to prevent the atoms in, say, your hand, from passing between the atoms in the cup, allowing you to touch but not penetrate the cup.

The weight you feel when you lift the cup, is created by the magnetic energy between the Quanta in the cup and those in the earth.

High-temperature scenario: If sufficient electro-magnetic energy (*heat*) is trapped by the electrons in the cup and your hand, the electric charge energy in all the atoms will exceed the magnetic field energy forcing the atoms in the cup and/or your hand to repel each other and intermingle, in a form that we understand as gas (Dalton's theory).

Because orbiting electrons cannot hold onto energy above a minimum level (e.g. microstates: N_t = 1.0; N_v = 1.5; N_p = 2.5), if an electron traps no further electro-magnetic energy, it will leave its proton partner and electro-magnetic radiation will cease from that proton-electron pair.

Chemistry, and therefore life, occurs when Quanta within a few atoms become detached or are gained through interaction. Thus, the creation of ions and isotopes causes electrical and magnetic disparity between adjacent atoms. Ions react electrically with other ions (chemistry) and isotopes create this capability.

And, finally, how do we perceive these energies:

Mass is the magnetic charge in atomic particles

Light is a particular range of electro-magnetic energy

Gravity is the potential energy between Quanta due to their magnetic charge.

Heat is the electro-magnetic energy emitted by a proton-electron pair

Temperature is the electro-magnetic energy emitted by the proton-electron pair(s) in an atom's innermost shell

All the energy in the universe is generated by friction within satellites by the competing potential and kinetic energies in their force-centres and sub-satellites, stored in neutrons and released during subsequent '*Big-Bangs*'.

All celestial bodies that are without either a force-centre or satellite (for example a galactic force-centre or moon) are black-bodies that generate no internal [heat] energy.

The entire universe comprises a fixed, unchanging quantity of energy, it always has done and always will. It was originally contained within the ultimate-body, released during the latest '*Big-Bang*' and remains unchanged today.
[first law of thermodynamics]

Electro-magnetic radiation is trapped by an electron, converted into kinetic energy, immediately lost to, and radiated by, its proton. If insufficient electro-magnetic energy is trapped by an electron, it naturally leaves its proton-electron partnership and continues in free-flight at the linear and angular velocities it had when it left its proton. However, electrons never stop moving.
[second law of thermodynamics]

The natural (minimum entropy) state of the universe is the reversion to protons and electrons
[third law of thermodynamics].

Electricity is the converse of magnetism and magnetism is the converse of electricity.
One cannot exist without the other.

2.1.1 **Electricity**

The application of potential energy (potential difference) sufficient to pull electrons from one atomic shell and transfer them to another along a conductor is what we currently refer to as DC (direct current) electricity.

Artificially induced electricity is generated by moving an electrical charge within a magnetic charge (e.g. in motors and generators) and is what we currently refer to as AC (alternating current – electrical field).

Electrical energy is shared between particles and travels from negative towards positive

2.1.1.1 Charge

The static electrical charge (e) can be negative (electrons) or positive (protons). It is ever-present in all Quanta and is all-pervasive. It radiates in all directions and its attraction/repulsion is felt by all other Quanta throughout the universe. It retains magnitude irrespective of distance.

It is a myth that the force induced by an electrical charge fades with the square of the distance from its source as this would conflict with the first law of thermodynamics. It is simply distributed over a larger spherical area.

Electrical charge is 2.269E+39 times magnetic charge, but because it is **shared** between Quanta:
the electrical attraction/repulsion between each of a million Quantum neighbours is one-millionth that of the electrical attraction between two Quanta. This is the reason we see negligible electrical attraction between celestial bodies (planets, stars, moons, etc.).

The magnitude of the electrical charge capacity (e or e_m) is defined by particle *mass* (m_e or m_p). The minimum magnitude (e) is always present and constant in all Quanta irrespective of circumstances.

A proton's additional mass provides it with the capacity to hold onto an additional electrical charge (e') commensurate with its electron's kinetic energy while it is part of a proton-electron pair. This additional charge is collected from its electron partner via the opposing static electrical charges and used by the proton to generate (and emit) electro-magnetic energy and electrical field energy.

The operational electrical charge (e') is the energy that repels adjacent atoms (gas).

When a proton loses its electron, its electrical charge will fall to that of the electron; the elementary charge unit (e).

According to Coulomb, electrical potential energy is calculated thus:
$PE = k.q_1.q_2/R$

2.1.1.2 Field

Electrical field energy is generated by proton-electron pairs and varies with proton electrical charge (e'; refer to Chapter 3.1.3). It acts as the carrier for the electro-magnetic energy radiated by a proton-electron pair.

Field electricity is artificially induced. It can be generated by bringing together two atoms (or collections of atoms) of strong opposite polarity (DC), or forced into a conductor by rotating a magnetic field (AC) such as passing an electrical conductor through a rotating magnetic charge (e.g. in motors and generators).

Field electricity is not addressed in detail in this book because it plays no part in Newton's laws of orbital motion.

2.1.2 Magnetism

Magnetic energy is accrued between particles and travels from positive towards negative.

Artificially induced magnetism is generated by moving a magnetic charge within an electrical charge (e.g. in transformers) and it is polar.

Magnetism is what we currently understand as mass and gravity (refer to Chapters 6.7 and 6.8 respectively).

2.1.2.1 Charge (Mass & Gravity)

Magnetic charge is non-polar magnetic energy present and constant in all Quanta. It is what we understand today as mass, and the field lines it radiates in all directions are what we understand as gravity.

It is a myth that the force induced by magnetic charge fades with the square of the distance from its source as this would conflict with the first law of thermodynamics. It is simply distributed over a larger spherical area.

Magnetic energy is 4.407E-40 times electrical energy, but because it is **accrued** between Quanta:
the magnetic strength of 100 Quanta is 100 times stronger than in one Quantum. This is why planets and stars - that comprise enormous numbers of Quanta - have such a strong magnetic attraction, but is so much weaker in, say, a cup.

The relative strength of the magnetic charge is defined by Quantum *mass* (m_e or m_p). It is always present and constant irrespective of circumstances. The magnetic charge in a proton is always m_p/m_e times greater than that of an electron.

Where proton-electron pairs can be aligned within a body due to their atomic lattice structures, they will produce strong magnetic polarisation. Whilst such alignment in iron (with a BCC lattice structure) must be induced artificially, e.g. with a lodestone, it occurs naturally in metals with an HCP structure such as Cobalt, Dysprosium, Gadolinium, Neodymium and Terbium. It is the magnetism that exists in bar-magnets.

Electro-magnetic radiation is deflected by magnetic charge.

According to Newton (and Gilbert), magnetic potential energy is calculated thus:
$PE = G.m_1.m_2/R$

2.1.2.2 Field

Magnetic field energy is generated by proton-electron pairs and unchanging. It is constant because the proton magnetic charge remains constant irrespective of electron energy. This is the energy that attracts adjacent atoms (viscous matter) and the magnetism that is generated by the earth.

Field magnetism is artificially induced. It is generated by encircling magnetic matter within an electrical charge (as in transformers). It generates polar field lines that flow from one end to the other.

This form of magnetism is much stronger than the magnetic charge, but it is highly concentrated, unidirectional and has a very limited field of influence.

Polar magnetic fields are highly selective in the materials they attract because of their dependency on nucleic alignment (lattice structure). Refer to Chapter 3.5.3 for numerical description of the nucleic and lattice (crystal) structure of metals.

Other than that generated by the earth, field magnetism is not addressed in detail in this book because it plays no part in Newton's laws of orbital motion.

2.1.3 Electro-Magnetic Energy

Electro-magnetic energy is generated by proton-electron pairs and is the means by which energy is transferred. It always travels at the same velocity (299792459 m/s), which we refer to as the *speed of light*, but is of course, the same speed for *all* electro-magnetic energy.

Electro-magnetic energy is susceptible to magnetic charge energy in a planet or star and will be deflected by it according to Newton's gravitational constant (G: refer to Chapter 6.2.1)

The bands of electro-magnetic energy (light, radio, X, γ, etc.) are defined by the orbiting electron's kinetic energy (velocity) and radiated in different wavelengths that we perceive as light, heat, etc. The magnitude of a radiated energy wave: its brightness, temperature, etc. is defined by its frequency. It possesses both electrical and magnetic polarised energy but no *mass* (refer to Chapter 6.2.2).

An electron can only collect (gain) from *electro-magnetic* energy greater than its own *kinetic* energy [second law of thermodynamics].

Electro-magnetic energy is emitted by a proton-electron pair in a direction normal to the plane of the orbiting electron according to the '*right-hand-rule*', and will travel in a straight line until deflected by magnetism, reflection or diffraction.

Orbiting electrons continuously collect electro-magnetic energy and convert it into kinetic energy (increasing their velocity). In doing so, proton-electron pairs generate (and radiate) electro-magnetic energy, simultaneously reducing kinetic energy in the electrons. This process of energy transfer continues only whilst orbiting electrons continue to collect electro-magnetic energy. I.e. it is:
1) collected by an electron
2) converted into kinetic energy
3) transferred between charges of opposite polarity (electron & its proton)
4) generates additional potential energy between the electron & its proton
5) collected by the proton using its additional electrical charge (e')
6) and simultaneously emitted by the proton

Calcium, for example, is an atom with twenty proton-electron pairs and will therefore emit electro-magnetic energy in twenty directions. Millions of such atoms will emit electro-magnetic energy in millions of directions.

Because the electro-magnetic energy radiated by the innermost proton-electron pairs within a body gets bounced around by its neighbouring pairs, it will take much longer to escape to atmosphere. This is why the surface of a body cools faster than its centre.

Electro-magnetic radiation is deflected as it passes a large body e.g. a planet or star because of non-polar magnetic charge, i.e. magnetism.

Heat is what we feel from the electro-magnetic energy emitted by proton-electron pairs. Temperature is how we measure this heat. Each atomic shell will emit heat at a different temperature, the highest being that emitted from the innermost shell and that which we measure (refer to Chapter 6.5.2). The key temperatures according to Planck are listed below (refer to Chapter 3.5.4.3):

T_o is Max Planck's minimum temperature
T_m is Max Planck's mean temperature
T_n is the temperature for neutron generation

Wave Characteristics

Electro-magnetic radiation travels in waves with a frequency, wavelength and amplitude commensurate with its energy. Its energy is what we understand (feel) as heat. Heat is the collection of energies radiated by proton-electron pairs, each of which will be at a different level dependent upon the orbital velocity of the electron responsible for it. The highest energy is emitted by a pair with its electron in the innermost shell and the lowest by a pair with its electron in the outermost shell.

In many publications, you will see the electro-magnetic wave depicted as described in Fig 1;**A**. The problem with this configuration is that the kinetic energy (E_T) of the electron must vary between 0 and 2.KE in each cycle for it to work, which is impossible. The kinetic energy in an orbiting electron remains constant throughout each cycle. The shape of the wave, therefore, must be as described in Fig 1;**B**, in which the total energy (E_T) in the wave is constant. I.e. when the electrical component is at a maximum amplitude, the magnetic amplitude is zero, and vice versa.

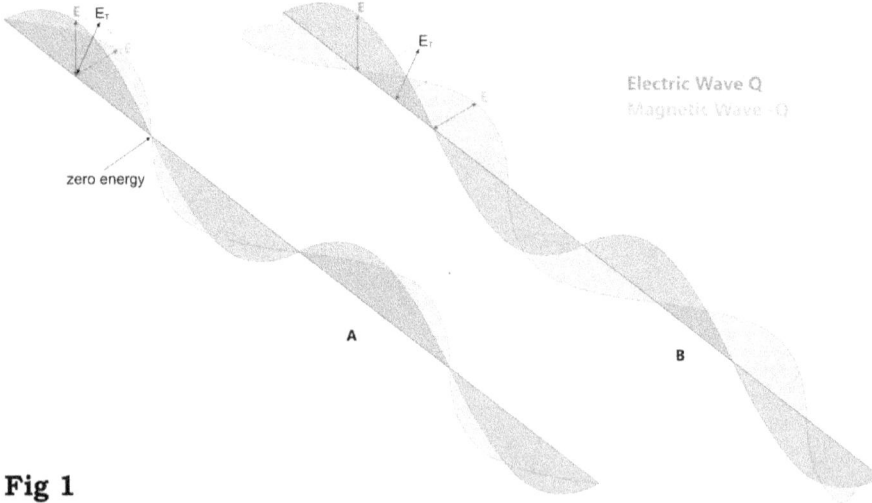

Fig 1

Nobody knows what an electro-magnetic wave actually looks like, so it could be similar to that shown in Fig 2, i.e. a helical wave that varies between maximum/minimum positive to maximum/minimum negative electrical and magnetic energy as it helically rotates. In fact, given the energy source (an orbiting electron), this option is more likely than that shown in Fig 1;**B**

Fig 2 shows a helical electro-magnetic wave winding its way through; $+E_e$; $-E_m$; $-E_e$; $+E_m$. Measuring either electrical or magnetic energy alone will produce a flat sinewave profile such as that shown, but it always emits the total kinetic energy in the electron (E_T).

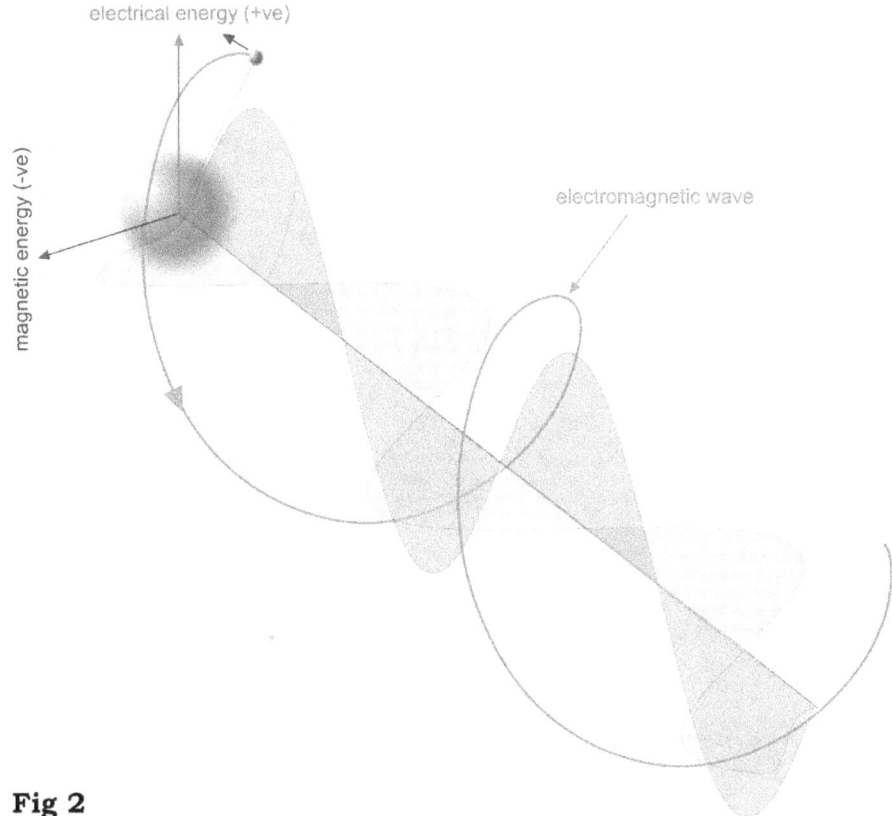

Fig 2

You will see many claims that amplitude increases with energy. This is incorrect. Whilst his *formula* (h) was incorrect, Max Plank realised that energy is proportional to frequency, not amplitude (refer to Chapter 6.11.5).

The electro-magnetic energy produced by each proton-electron pair will be different to that produced by a proton-electron pair in the same atom but in a different shell. The difference will be consistent with the kinetic energy in the electron generating the energy and is what gives the different atomic numbers their characteristic colour (light) banding (Balmer lines).

Irrespective of wave shape, electro-magnetic energy possesses the [kinetic] energy transferred by the orbiting electron and is emitted by the dynamic electrical charge held by the proton (e'), which varies between electrical and magnetic energy according to equal sinewave profiles normal to each other.

2.1.4 Potential Energy

Potential (static) energy exists between all Quanta, irrespective of separation distance in the form of what we currently understand as gravity but is actually due to the combined magnetic charge in all the Quanta in any and all bodies.

Potential energy is the attraction or repulsion of either electrical or magnetic energy.

Negative potential energy (PE: gravity) holds particles together and positive potential energy (CE: centrifugal) keeps them apart.

In a balanced system (e.g. orbits), both PE and CE must be equal.

2.1.5 Kinetic Energy

Kinetic (dynamic) energy, which exists in all moving particles, is always positive [7] and transferred via electrical, magnetic, electro-magnetic or impact [potential] energy.

Electro-magnetic energy is absorbed by orbiting electrons and converted into kinetic energy.

Kinetic energy in a satellite following a circular orbit (such as in an atom) is not induced into the satellite by its force-centre such as in elliptical orbits; it must be provided by the satellite itself.

2.1.6 Heat & Temperature

Heat is the combined electro-magnetic energy emitted by all the proton-electron pairs in all the atoms in matter.

Temperature is the electro-magnetic energy emitted by the proton-electron pairs whose electrons are orbiting in the innermost shell of the atoms in matter.

All heat is radiated:
Conduction is the transfer of electro-magnetic energy radiated between adjacent atoms in viscous matter (solid/liquid).
Convection is the movement of atoms with high electrical field energy (repulsion) to a position where this energy can balance in three-dimensions, i.e. where its neighbouring atoms possess less energy. The high-energy atoms will subsequently transfer [radiate] electro-magnetic energy to the cooler neighbouring atoms (second law of thermodynamics) until balance occurs. It only occurs in gases that are under the influence of magnetism (gravity).

Refer to Chapter 4.5 for the properties of key temperatures associated with electro-magnetic radiation.

2.2 Orbits

In an effort to avoid confusion, I shall use the terms *mass* and *gravity* to describe bodies and their mutual attraction instead of the more accurate term magnetism.

An orbital system comprises one or more satellites orbiting a force-centre.

If Newton's laws are valid for one orbital system, e.g. a solar system, it is reasonable to presume that they apply to all other orbital systems, including the atom. Alternatively, it is unreasonable to claim that a law proven to apply to one orbital system does not apply to all of them.

Therefore, it must hold true that Isaac Newton's laws *must* be universal, exactly as he claimed. So, they must apply to all naturally occurring orbital systems, including atoms. To demonstrate this, I have successfully applied Newton's laws to all orbital systems from atoms to galaxies. Proof of Newton's laws of orbital motion is provided in Chapter 6.1

The laws considered here apply to orbital systems that comprise force-centres and satellites that can be spiral galaxies, solar and lunar systems, and atoms; spiral galaxies being the largest and atoms the smallest. A galactic orbital system comprises a force-centre about which its satellites (stars) orbit; planets orbit the stars and moons orbit the planets.

A force-centre can be a proton, planet, star or central body of a spiral galaxy about which all its satellites orbit.

A satellite can be an electron, a moon, a planet or a star in a spiral galaxy that naturally orbits its force-centre.

An orbit is the elliptical path traced by a satellite around its force-centre. Particle spin (refer to Chapter 2.3) plays no part in this theory.

An orbital path is always an ellipse that may be flat ($e < 1$) or round ($e \geq 0$), e.g. a circular orbit ($e = 0$)

It works like this:

A planet's orbital motion is maintained via the gravitational acceleration induced between it and its force-centre from its apogee, and subsequent deceleration induced by the same force as it passes its perigee.

Because of the conservation of energy and the fact that the orbit is a symmetrical ellipse, the potential & kinetic energies on both sides of its orbital principal axis are equal and opposite.

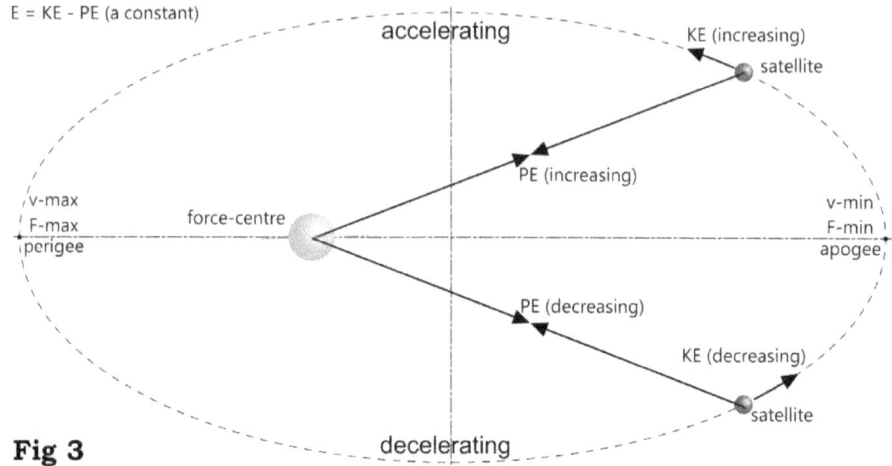

Fig 3

Fig 3 shows how the elliptical orbit is the perfect, perpetual motion machine, the orbit never changes (refer to Chapter 2.2.8), but it only works in a vacuum. As a satellite accelerates, both its KE and PE increase. But as KE is positive and PE is negative, total energy (E = KE + PE) remains constant, ensuring the conservation of energy:

Not only is there no need for the distortion of space-time and gravity (Relativism) around a force-centre to explain satellite performance, such a theory actually requires the contradiction of a fundamental law of nature (elliptical orbits) in order to function (refer to Chapter 6.2.4).

Isaac Newton described the coupling forces, and therefore the potential energies, involved in keeping an orbiting satellite in motion around its force-centre thus:

$F = G.m_1.m_2 / R^2$
$PE = G.m_1.m_2 / R$
Where:
G is Newton's gravitational constant
m_1 represents the mass of the force-centre (e.g. a star)
m_2 represents the mass of the satellite (e.g. a planet)
R represents the distance between the centres of m_1 and m_2
F represents the gravitational force acting between m_1 and m_2
PE represents the potential energy acting between m_1 and m_2

Strictly speaking, Isaac Newton's formula is not quite correct, and neither is his constant:
G = 6.67359232004334E-11 m³ / kg.s²
It should read:
$F = G.m_1.m_2 / A$ and 'G' should equal 8.3862834423E-10 m³ / kg.s²
Whilst this does not alter his formula, it does focus the mind on the relationship between gravitational force (or energy) and distance. It is currently, and incorrectly, believed that gravitational force diminishes with the square of the distance from the centre of its source. If this were true, the law of conservation of energy would be violated. Gravitational force is the same at any distance from the centre of its source, however, at any given radius it must be distributed over the spherical area at that radius, which is why it *appears* to diminish with the square of the distance from the source. It works in the same way as radiated light.

Whilst the above formula describes Isaac Newton's universal law of gravity, it is more useful in the form of potential (gravitational) energy, which we can achieve by the simple expedient of multiplying the force by the separation distance (R):
$PE = F.R = G.m_1.m_2 / 4.\pi.R$ (using the corrected value for 'G' above)

From this simple expedient we can complete Newton's laws insofar as they apply to orbiting satellites.

A satellite's velocity varies around its orbital path from a maximum at its perigee (or perihelion) to a minimum at its apogee (or aphelion). The terms perigee and apogee are normally applied to lunar orbits, and perihelion and aphelion are normally used for planetary orbits, but they mean the same thing.
The perigee is the point at which a satellite's orbit passes closest to its force-centre.
The apogee is the point at which a satellite's orbit passes furthest from its force-centre.

Orbital accuracy can be established by calculating the orbital apogee radius using both orbital dimension and system energy. If correct, this radius will be *exactly* the same in both calculations. This verification technique has made it possible to demonstrate that the angular kinetic energy of a satellite plays no part in Newton's mathematical model.

Isaac Newton was clearly a little paranoid about giving away his discoveries, so he hid a number of crucial details making his theories somewhat difficult to interpret exactly and therefore to prove and use. This may account for the emergence of various scientific myths (refer to Chapter 2.7). It is a pity that Newton's paranoia influenced his actions because science might otherwise be a lot further ahead. In fact, Isaac Newton's achievements were greater than even he could have expected:

Newton's theory of orbital motion has been successfully applied to all orbital systems from atoms to galaxies, thereby proving that the concept of sub-atomic dark matter is not required to explain the sun's orbit in the Milky Way.

Using Newton's laws, it has been possible to prove that any two satellites may be swapped without altering the shape or period of the respective orbits; e.g. Earth with Jupiter, Mercury with Halley's Comet, etc. The only orbital variations will be kinetic and potential energies.

Newton's formula $F = G.m_1.m_2 / R^2$ also allows us to determine [core] pressure anywhere inside a solid body.

The potential and kinetic energies coming from Newton's laws provide everything needed to calculate the angular (spin) kinetic energy in satellites and force-centres and thereby determine their polar moments of inertia that together with Core Pressure, allows us to estimate their internal structure.

Together with Newton's laws of orbital motion, spin theory has shown us where all universal energy is generated.

His laws have allowed us to prove that the potential energy in a proton-electron pair is exactly twice the electron's kinetic energy and thereby predict the creation of a neutron and the true meaning of $E=mc^2$

Newton's gravitational constant (G), which is based upon the properties of Quanta, allows us to calculate the properties of atoms and to predict the deflection of light passing celestial bodies (refer to Chapter 6.2.1)

The reason massless electro-magnetic radiation can be deflected by a planet is because gravity is actually magnetism.

Together with spin theory, Newton's laws of orbital motion have enabled us to define the size and spin rate of Hades (refer to Chapter 3.3.6) along with its position in the Milky Way.

Between them, Newton and Kepler resolved the single most important feature of natural physics in the universe.

The following chapters describe the mathematical principles that define the behaviour of satellites in their orbital paths and their spin characteristics. Don't worry, they are very simple. However, you need to remember a few things when reading them.

Terminology

An **orbit** is the path followed by a satellite around its force-centre. For example, our moon is in orbit around our Earth (a planet), which is in orbit around our sun (a star), which is in orbit around Hades (a galactic force-centre). Electrons orbit their protons (a proton-electron pair).

An orbiting body (or mass) is referred to as a **satellite** and the body about which it orbits is referred to as its **force-centre**.

These orbital systems have group names such as:

Solar Orbit: A star's orbital path around its galactic force-centre
Planetary Orbit: A planet's orbital path around its star
Lunar Orbit: A moon's orbital path around its planet
Atomic Orbit: An electron's orbital path around its proton

Collectively, everything (including the force-centre) orbiting a ...
... galactic force-centre is called a **galaxy**
... star is called a **solar system**
... planet is called a **lunar system**
... proton is called a **proton-electron pair**

An orbit is *always* a perfect ellipse, exactly as Johannes Kepler stated. An ellipse can be any two-dimensional (flat) elliptical shape including a circle.

There is a major difference between a genuine elliptical orbit, i.e. one in which its axes *are not* identical in length, and a circular orbit, i.e. one in which both its axes *are* identical in length:

Satellites following **elliptical orbits** (e.g. stars, planets, moons, comets, etc.) keep going because of the potential and kinetic energies (Newton's constant of motion) between a satellite and its force-centre.

Satellites following **circular orbits** (e.g. electrons, communication, spy, etc.), keep going because they provide their own kinetic energy.

Mass is magnetic charge

Gravity is the potential energy generated by magnetic charge

Velocity in orbits refers only to the curvilinear motion of a satellite in its path around its force-centre. It does not refer to rotational (angular) motion in a satellite or its force-centre

Force is energy per unit distance

Energy is a force applied over a specified distance

Kinetic energy is the energy in a satellite due to its velocity

Potential energy is the attractive/repulsive energy between a satellite and its force-centre

Planetary Spin; the angular velocity (radians per second) in a body rotating about an axis that passes through its centre of mass

2.2.1 The Laws

The laws of orbital motion are the mathematical formulas that describe the properties of a satellite's curvilinear motion around its force-centre. It is important to understand that rotary motion (spin) plays no part in Isaac Newton's laws of orbital motion. However, the laws describing spin in a satellite and its force-centre may be derived from them.

Newton's laws of orbital motion for non-circular elliptical orbits come in two parts:
1) orbital dimensions and period
2) potential and kinetic forces (and energies)

The calculation results from one cannot be used to validate the other because the force-centre *alone* defines orbital shape and period. I.e. it is possible (in theory and in practice) to swap any satellite from its own orbit with that of another without altering the orbital periods or shapes. The only differences will be the orbital forces and energies.

In other words; whilst you can calculate the mass of a force-centre from the dimensional shape of any one of its satellite orbits, you cannot calculate satellite mass.

A satellite can only have one force-centre; once a satellite and force-centre have paired-up, the union cannot be broken unless by force. Even in the case of a binary star (refer to Chapter 2.3.6), each satellite remains tied to its original force-centre. This is the reason every electron in every atom remains *attached* to its proton partner, irrespective of the presence of all the other protons in the same nucleus.

A satellite's orbit (shape and period) is not altered by its own orbiting secondary satellites (e.g. moons).

The most important of all orbital laws is:

Every orbital system must have one force-centre and at least one satellite

I.e. a satellite cannot exist as such without a force-centre (therefore; Hades *must* exist)

2.2.2 Elliptical

Newton used the orbits in our solar system to prove his laws of motion, which gradually became universally accepted until the early 20th century. It looked something like that shown in Fig 4 below:

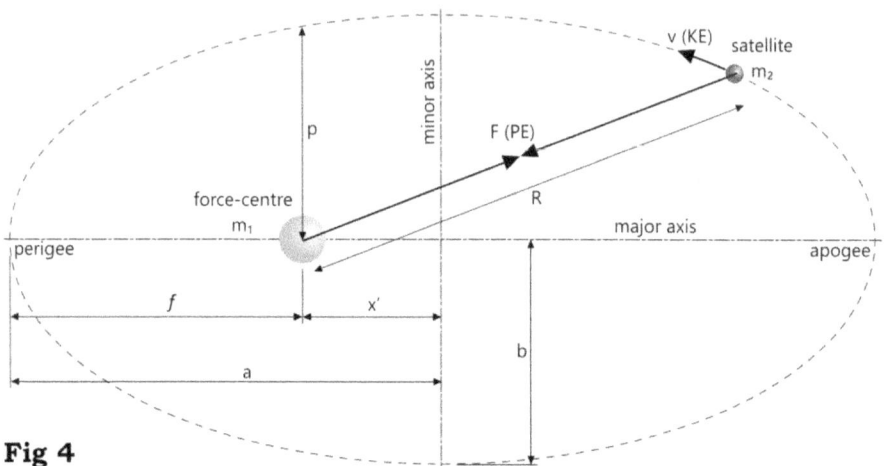

Fig 4

A solar system is essentially a star with its orbiting planets that have their own orbiting moons. Whilst Newton's laws define the orbital properties of a satellite, they do not establish its *angular* velocity. However, the potential and kinetic energies resulting from Newton's laws do provide the information required to solve this problem and therefore to estimate its (the satellite's) internal construction. This means that virtually everything we need to know about our solar system can be determined from Newton's laws of orbital motion.

Newton's constant of motion (h) keeps the satellite moving (refer to Chapter 6.1.4.3).

The shape of an orbit is an ellipse (thank you Kepler!) with an eccentricity between 0 and less than 1. The force-centre is always located at its focal point, which is on the major axis and the closest point at which an orbiting satellite passes its force-centre. A satellite's acceleration as it approaches its force-centre and deceleration as it passes in the opposite direction is what keeps it going (Fig 3).

A useful tip from Kepler that was later verified by Newton is that the relationship between the swept area inside the ellipse for any given period of time of a satellite's orbit will always be identical (Fig 5).

PHILOSOPHIÆ NATURALIS PRINCIPIA MATHEMATICA Revision IV

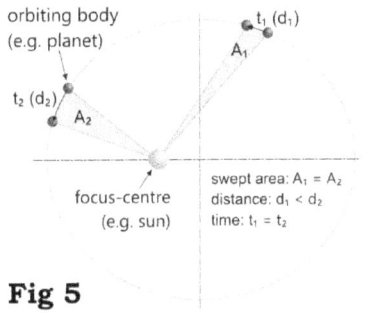

Fig 5

171 solar system orbits have been reproduced using Newton's laws in the creation of this book, all of which have been verified using data published by NASA; *no exceptions*:

So, we can confidently state that Newton's laws *are* universal.

2.2.3 Circular

The parameters for a circular orbit are the same as those described for elliptical orbits (refer to Chapter 2.2.2), with the following exceptions:

Satellites following a circular path such as in the atom and man-made satellites must provide their own kinetic energy. In the case of electrons, electro-magnetic energy is absorbed and converted into kinetic energy [6].

The potential energy between a force-centre and its satellite following a *circular* orbit is *always* twice the kinetic energy in the satellite.

Because a satellite's potential energy increases with increasing velocity (kinetic energy), its orbital radius also decreases. Therefore, atomic strength increases with increasing temperature.

2.2.4 Force-Centre Mass

The mass of any force-centre can be calculated accurately from just two geometric properties of one of its satellites.

If you know half the distance between a satellite's orbital apogee and its perigee (half the length of its major axis) and its orbital period, you can calculate its constant of proportionality (K; refer to Chapter 3.2.4). And this constant can be used to calculate the force-centre's mass.

You can use 'K' to determine the dimensional properties of all the other satellites orbiting the same force-centre knowing only their perigee (or apogee) distances.

$m_1 = (2\pi)^2 / G.K$

2.2.5 Planetary Mass

If a satellite's mass can't be determined simply from its orbit and its velocity within it (refer to Chapter 2.2.1), how do you calculate the mass of everything in the solar system with only a ball of known mass, a watch and observed orbital data?

Galileo kicked us off with his discovery of gravitational acceleration, albeit his attempts to find an accurate value were hampered by his inability to measure time accurately.

By dropping our ball (or rolling it down a slope) and measuring the time it takes to cover a known height, we can find the gravitational acceleration on the surface of our planet, an accurate value for which is now available (refer to Chapter 2.2.5.1).

You can use this value (g) along with the measured gravitational force of a known mass at a planet's surface to give you the planet's mass. This method can only be used, however, if you have access to the planet's surface.

Alternatively, where such access is unavailable, you can use a planet's deviation from its known orbital path to calculate its mass (refer to Chapter 3.2.5).

Together with a planet's mass, you can use its angular velocity to estimate its internal composition (refer to Chapter 2.4.1).

2.2.5.1 g

An *'ISO'* value for the gravitational acceleration on earth's surface is described below:

a) ISO states that; 1lb (force) = 4.448222N (exact)

b) ISO also states that; 1lb (mass) = 0.45359327kg (exact)

c) The value for 'g' appears if you divide 1lb (force) by 1lb (mass)
   g = 4.448222 ÷ 0.45359327

d) g = 9.80663139027614 m/s²

... which happens to be the value at sea-level at latitude 45.5° (Milan or Minneapolis).

Because we know the radius of our planet at 45.5° (6371000.685 m), we can calculate the mass of the earth quite accurately;
g = G.m / R²
m = g.R² / G = 5.95786303763712E+24 kg {m/s² . m² . kg.s²/m³}

2.2.6 External Influences

Satellites are often but temporarily influenced by other celestial bodies. For example, Fig 6 shows a planet orbiting its sun but close to another body outside its orbit. The gravitational energy between the satellite and the external body will pull them together, out of their respective orbits temporarily altering their orbital paths. The satellite will accelerate as it travels towards the external body and decelerate (relative to the external influence) after passing it. However, these variations are effectively cancelled out as a result of Kepler's and Newton's 'equal time swept-area' law and the conservation of energy.

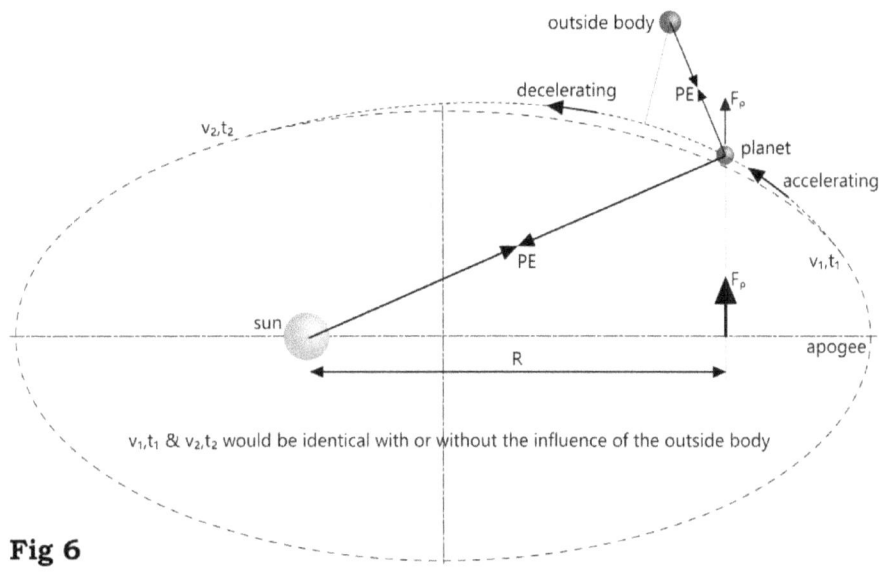

Fig 6

Whilst the orbital period is maintained, these temporary influences apply a torque ($T_p = F_p \times R$) to the orbital axes causing a gradual rotation (of the orbit) about its force-centre. The frequency and magnitude of such influences define the rate of orbital precession.

These deviations may be used to determine the mass of one body if the mass of the other is known (refer to Chapter 3.2.5) using the *'triangle of forces'* calculation method.

2.2.7 Centrifugal Force

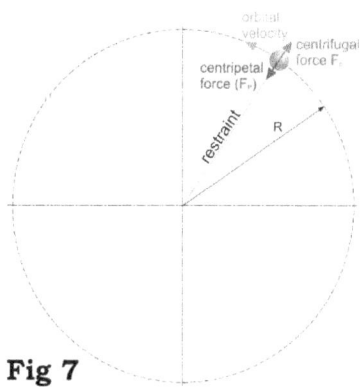

Fig 7

Any orbiting mass will be subject to a centrifugal acceleration (a) that must always balance with the acceleration (g) induced by gravity.

If you swing a ball - tied to a length of string - around your head, *centrifugal* force is pulling the ball away from you. But it also induces a tensile force in the string, pulling the ball towards you. This is *centripetal* force, but it is also *potential energy*; the equivalent of gravity.

Christiaan Huygens gave us the mathematical relationship between this and its velocity in a circular orbit (Fig 7); $a = v^2 / R$, where 'v' is its curvilinear velocity and 'R' is its orbital radius.

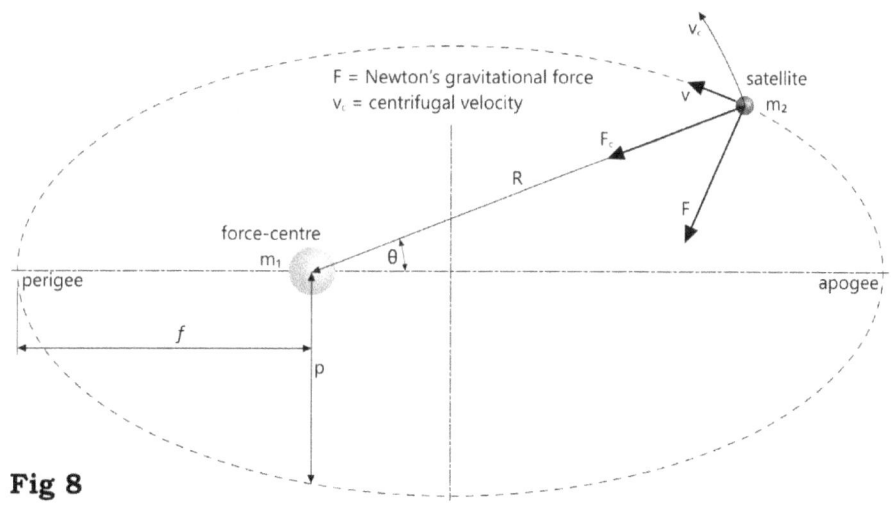

Fig 8

However, the above velocity (v) must be modified for elliptical orbits (v_c) as shown in Fig 8. Its magnitude is dependent upon the orbital eccentricity and varies with orbital radius (R).

Refer to Chapter 3.2.7 for the formulas that will enable you to calculate the centrifugal force in an elliptical orbit.

2.2.8 Station-Keeping

When a displacement force pulls a satellite off course, a restoring force will ensure that after release the satellite will return *exactly* to its orbital path, where both centrifugal and gravitational acceleration balance; i.e. where they are equal and opposite.

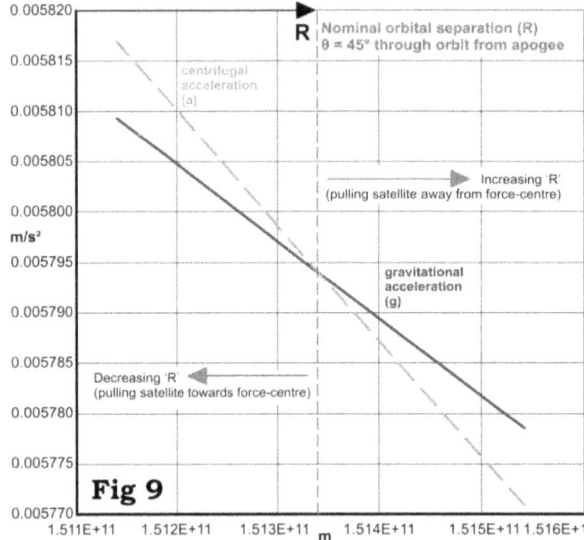

Fig 9

Fig 9 provides a graphical representation of the restoration force on the earth 45° through its orbital path from its apogee.

As the earth is pulled away from the sun (increasing R), gravitational acceleration (g) increases faster than centrifugal acceleration (a), pulling the earth back towards the orbital path when the displacement force is released.

As Earth is pulled towards the sun (decreasing R), centrifugal acceleration (a) increases faster than gravitational acceleration (g), pulling the earth back towards the orbital path when the displacement force is released.

As you can see (Fig 9), *exact* balance occurs at the orbital path: at nominal orbital separation 'R'.

This relationship between centrifugal and gravitational acceleration is what maintains the orbital path in the event a satellite comes under the influence of external forces (refer to Chapter 2.2.6).

The above process requires a perfect elliptical orbit, *exactly* as Kepler defined (refer to Chapter 6.2.4).

2.2.9 Orbital Planes

The spin (refer to Chapter 2.3) induced in a force-centre by its orbiting satellites, influences the orbital plane of its satellites (Fig 10).

Fig 10

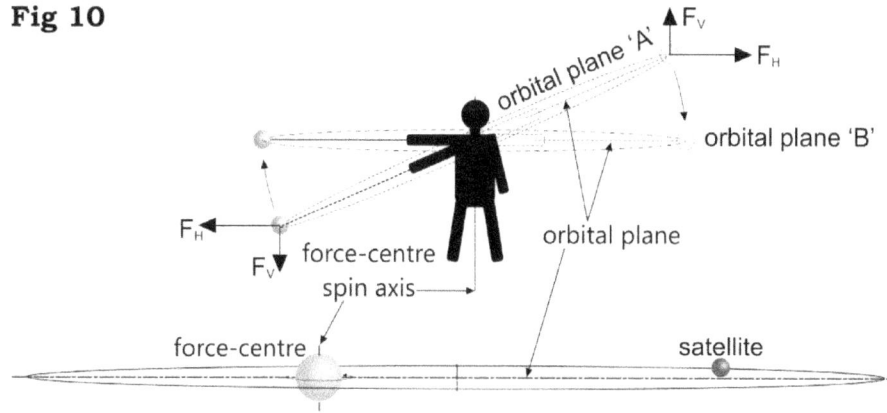

The rotational kinetic energy in a force-centre naturally causes the orbital planes of its satellite's to settle at 90° to its spin axis. This phenomenon can be demonstrated by attempting to swing a ball about *orbital plane 'A'* in which 'Fv' is non-zero. The orbital plane will always settle at the lowest energy condition; *orbital plane 'B'*, where 'Fv' is zero. The same forces are at work in planetary orbital systems.

2.2.10 The Importance Of Orbits

Everything in the universe, without exception, is dependent upon orbits.

Without them, there would be no heat, elements, chemistry, life, electromagnetic energy or matter. In fact, the entire universe would be a cold, dark sea of Quanta.

Orbits are the source of *all* universal energy; through spin (friction) and neutrons.

So, it is not surprising that Newton's laws of orbital motion are without doubt *the* most important mathematical laws in science. They explain how *everything* in the universe works, including the atom!

Without the electron orbit, there would be no atoms. There would be no elements and therefore no molecules. Neutrons would not exist. Electrons would vanish into infinity, leaving behind a sea of lone protons.

If you were to remove the lunar orbit, e.g. the earth's; there would be no differential core-mantle rotation and therefore, no planetary heat or magnetic field. Days would be excessively long (one day would be longer than a year). There would be no continental drift, no atmosphere, no tides, no weather, no volcanic activity, etc., and therefore no life as we know it.

If we were to lose the planetary orbit, we would have no seasons, no rotating sun; one face of the planet would remain permanently hot whilst the other would remain permanently cold.

If we lost the galactic orbit, our sun would not be generating its internal heat or light; i.e. there would be no stars and hence no neutrons and therefore no subsequent 'Big-Bang'!

2.3 Spin

Spin is the source of all universal energy through internal friction.

All astrophysicists today claim that either there is no force spinning the planets or that it is impossible to calculate owing to the chaotic nature of the solar system. Both these views are incorrect. Planetary spin can be easily predicted with the same degree of accuracy as Newton's laws of orbital motion from the potential and kinetic energies calculated using them. This same theory also applies to galaxies and the Atom.

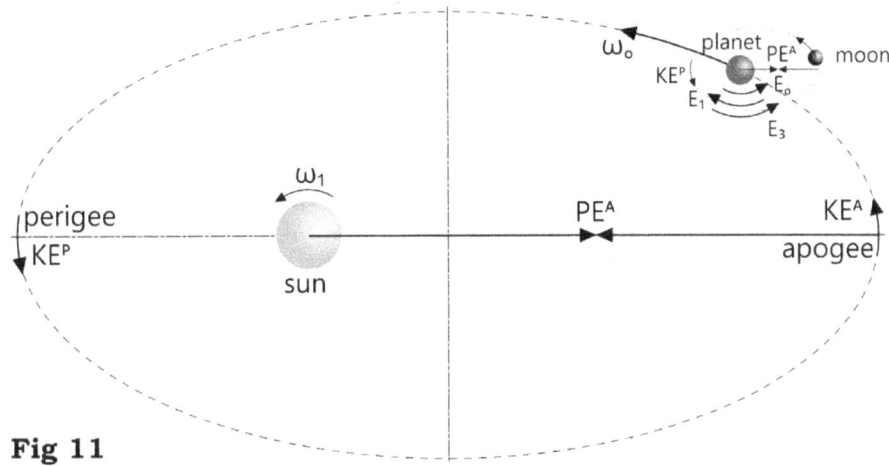

Fig 11

Fig 11: Potential energy between a force-centre and that of its satellite will naturally cause the satellite to spin at the same angular velocity and in the same rotational direction (prograde) as its orbit around the force-centre (ω_0). However, spin energy induced in the satellite by the force-centre's own rotational kinetic energy will cause the satellite to rotate in the opposite (retrograde) direction ($-\omega_1$).

If a planet's moon is orbiting in the same rotational direction as the planet's orbit e.g. prograde, the moon's kinetic and potential energies will cause the planet to rotate also in a prograde direction (ω_1).

Only the gravitational and kinetic energies in a force centre, its orbiting satellite(s) and their secondary satellites induce spin in each other. It is not as difficult or complex as everybody appears to believe.

For example: Venus's opposite rotational direction cannot, apparently, be explained, but the reason for it is quite simple once you understand spin theory.

E_3, which is generated by a planet's moon(s), is always considerably greater than E_1 and E_0 (Fig 11) and therefore defines a planet's rotational direction. As E_3 in a planet with no moon is zero its orbital energies (E_1 & E_0) determine its spin direction. Being a (relatively) large planet, E_1 is the dominant factor in Venus's spin direction.

The same argument applies to Mercury except in its case its smaller size means that E_0 is dominant so it spins in the same direction as the other planets in our solar system.

A significant difference will occur between the angular velocity of a planet's core and its mantle only if it has a substantial moon.

The only unknowns you need in order to calculate planetary spin are the polar moment of inertia of the planet and that of its sun. When calculating the spin in a planet, you need either its angular velocity (ω) or its radial modifier (Δ). If you have one, you can calculate the other.

The competing rotational energies (E_1, E_3 & E_0) along with their different distribution (@ the core or throughout) are responsible for the conflicting angular velocities in a planet's core and mantle and thereby generating its internal heat. This velocity difference in the earth (refer to Chapter 2.3.2) is also responsible for generating the earth's magnetic field. The positive value of $\delta\omega$ in the earth means that the right-hand rule depicts magnetic north in the correct direction.

Gas planets exist as such because they have been able to attract sufficient satellite *mass* to melt their [surface] crusts through internal friction. Unlike a star, however, they cannot generate the heat required to create neutrons. It is also probable that the heat lost to a gas planet's heavy surface gases will form a surface skin.

This calculation method is as accurate as Newton's own laws of motion and is essentially an extension of them. Therefore, not only is planetary spin predictable, it is both simple and accurate.

Using this calculation method, it has been possible to determine what would happen to the earth's spin if it lost its moon (refer to Chapter 2.3.5).

PHILOSOPHIÆ NATURALIS PRINCIPIA MATHEMATICA Revision IV

The component spin energies in a celestial body are summarised below:

E_0 is the natural rotational energy developed in a satellite, assuming it presents the same face to its force-centre. It may be represented by the rotation that would be expected in a ball swung about your head, at the end of a length of string. The direction of spin induced in a satellite by 'E_0' will be the same as that of the satellite's orbit. The spin-period will be identical to the satellite's orbital period.

E_1 is the rotational energy induced in a satellite by the rotational energy in its force-centre. This rotation may be represented by a pair of gears and behaves in the same way; i.e. the direction of spin induced in the satellite by 'E_1' will be opposite to that of the satellite's orbit.

E_2 is the final (total) spin in the satellite; i.e. the sum of all the other rotational energies.

E_3 is the spin energy induced in a satellite by its own sub-satellite(s). The direction of spin induced in the satellite by 'E_3' will be the same as the orbital direction and kinetic energies in these sub-satellites.

2.3.1 Polar Moment of Inertia

Polar moment of inertia of an homogeneous mass is calculated thus:
$J = ⅖.m.r^2$
Where 'm' is the mass of the satellite and 'r' is its outside radius.

However, this formula only works as shown if the body comprises a mass of constant density, which is not the case for celestial bodies, such as; planets, stars, moons, etc. Gravity tends to ensure that the denser matter migrates towards their cores.

This problem can be solved by using a radial modifier thus:
$J = ²/_5.m(Δ.r)^2$
the radial modifier (Δ) defines a body's radial centre of mass according to its variable density.

If you know a body's angular velocity (ω) you can calculate its radial modifier (Δ). Alternatively, if you know its radial modifier, you can calculate its angular velocity.

When used in conjunction with Core Pressure (refer to Chapter 2.4), 'Δ' can help us to determine a body's internal structure.

2.3.2 Earth's Core

The earth's core is a large ball of [mostly] iron, the spin of which is controlled by the potential energy of the earth's force-centre (its sun).

Spin in the earth's mantle, however, is dominated by the kinetic and potential energies of its moon. Both of these energies, which are driving spin in opposite directions, are creating internal friction between its core and its mantle.

Rotation of the earth's core ($\approx$7E-05 radians per second relative to its mantle; refer to Chapter 3.3.2) generates its magnetic field and the friction that is the source of the earth's internal heat.

Internal frictional heat is the source of all universal (electro-magnetic) energy, including the stars.

Here's an interesting question! *What would happen to the earth if it acquired another significant moon?*

2.3.3 Earth's Magnetic Field

The earth's lunar tilt angle (Fig 12: α = 23.4°) is a clear indication that the earth acquired its moon from outside its solar system (galactic comets), as is the case for all of the moons in our solar system, and perhaps, most of its planets.

Spin in the earth's core is dominated by our sun and spin in its mantle is dominated by its moon. The resultant relative spin rate, together with the magnetic properties of what we understand today as *mass*, in the earth's iron-rich core and mantle are responsible for generating the earth's polar magnetic field.

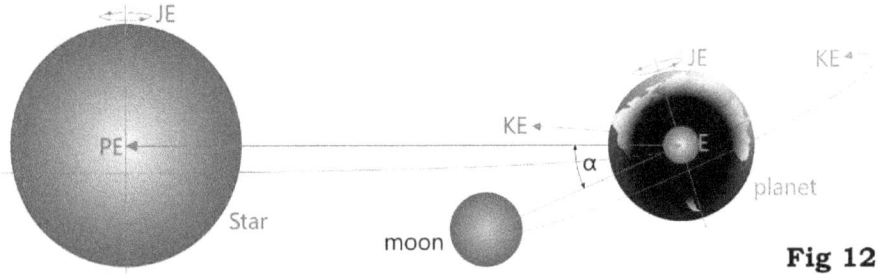

Fig 12

If a planet's lunar orbital plane is not coincident with its solar orbital plane, its mantle and core will spin on different axes, generating an angular difference between its true (physical) North and its magnetic North (6.277°; refer to Chapter 3.3.3)

This phenomenon is confirmed by; the lack of a magnetic field in Venus and Mercury, and the massive magnetic field in Jupiter that acts at 90° to the orbital plane of its moons, similar to the earth. In fact all planets and stars with satellites will generate a magnetic field ≈normal to the orbital plane of their satellites.

2.3.4 Magnetic Reversal

A worrying aspect of the earth's magnetic field is that the relative spin induced in the earth and its core by our moon and our sun will not reverse unless either the earth or its moon changes orbital direction, which is highly unlikely. Something external to the earth (extra-terrestrial) must therefore cause this reversal to occur periodically.

It would appear that the earth's magnetic reversal can only be explained by flipping it through 180°; switching north and south poles!

We already know that our solar system has a number of orbiting comets, so it is highly likely that the Milky Way also has its own 'comets', and these could be planet sized. Therefore, a large galactic comet may well be responsible for flipping the earth and/or any other planet in the solar system as it passes close by.

2.3.5 No Moon!

If the earth lost its moon, it would also lose its tilt (and its seasons), there would be insufficient internal heat energy to drive its continental plates and its magnetic field would vanish. The loss of internal friction would cause the earth's surface temperature to fall by about 200K.

The earth would therefore behave similarly to Venus except for its atmosphere, most of which would liquefy/solidify because the earth' surface receives only 52% of the sun's radiated heat compared to Venus.

Without its moon, one earth day would be 12450.152 hours (<519 current earth days) and the sun would rise in the West and set in the East just like Venus. There would be no seasons because the earth would lose its tilt.

2.3.6 Goodricke & Algol

In 1784, John Goodricke discovered the [supposed] binary nature of the star originally named Algol. As one of the binary stars passed in front of the other, their combined brightness dimmed, revealing two important facts:

1) One of the stars was bright (hot) and the other was dark (cold)

2) Only the bright star was a force-centre for an orbital system

The heat generated by the *bright* star is due to its dedicated satellite population. The other darker star (or large planet) is actually in orbit around the bright-star but has no satellites of its own. In which case, the dark partner can generate no internal heat.

Once a star (force-centre) has acquired its satellites, it can trap a twin but it will not share its satellites.

This discovery also reinforces the fact that stars generate their own internal heat from their satellite population and the proton-electron pair behaviour in atoms.

2.3.7 Chicken & Egg

Which came first; spin or orbit?

What you see in most films and documentaries is that the sun starts spinning and the planets follow it around. This is of course 'back-to-front'.

In order to generate spin, you need an appropriate energy. Spin theory teaches us that if a sun, planet or moon sat alone in space it would not spin.

Spin in our sun was first induced by the rotational energy in its force-centre (Hades) and it would have continued to spin at this rate had it not acquired its satellites (planets). However, our sun actually rotates at more than ten times this speed.

If angular kinetic energy in a force-centre induced orbital kinetic energy in its satellites, this transfer of energy would slow down the force-centre's rotation, which is obviously not the case. I.e. kinetic energy in the planets can and does induce rotational kinetic energy in the sun.

Therefore, the planets must have been orbiting long before the sun achieved rotation anywhere near its current rate.

The same argument applies to a spiral galaxy. Our sun got its initial spin (<2E-07 radians per second) from the spin energy in Hades, but Hades has no force-centre. Therefore, all of Hades spin comes from its orbiting satellites.

So, orbits came first!

2.4 Core-Pressure

This astro-physical property appears to have been overlooked to date but can be readily defined using Newton's revised formula: $p = G.m_1.m_2/A^2$ in which m_1 is the mass inside radius 'r' and m_2 is the mass outside 'r' and 'A' is the spherical surface area at radius 'r'.

An example calculation that yields a surprising, but understandable result has been performed for the earth (Refer to Chapter 3.4).

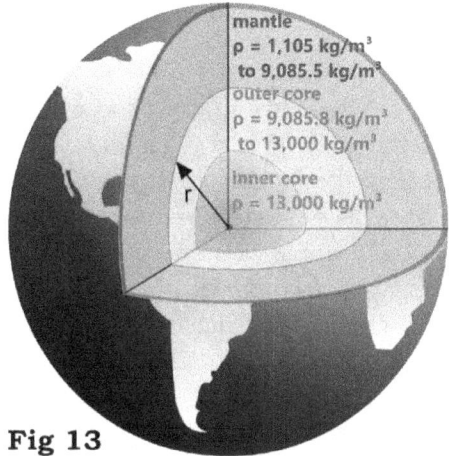

Fig 13

Most publicly available sources claim that the earth has a core density of somewhere around 13000 kg/m³, which, given the incompressibility of iron, appears unlikely unless it contains a much larger percentage of heavy metals than is currently claimed. However, whilst the structure described in Fig 13 provides the correct total mass and polar moment of inertia for the earth, because of its relatively low percentage mass, reducing the core density will have little effect on the density of the upper mantle material (immediately below the earth's crust) that cannot be much more than that of water.

Whilst this discovery may be unexpected, it is perfectly logical. It is currently claimed that mountain roots beneath the earth's crust eventually fall into the upper mantle, causing the crust to rise locally and temporarily. This event would be difficult to explain if the density of the mantle were greater than, or even close to, that of the crust. It is not so difficult to understand, however, if the upper mantle is a hot, gaseous, pressurized, cauldron of matter with a lower density than the crust it is supporting. Moreover, it is also easier to see where all the volcanic activity comes from. The same applies to subduction zones where crust material is pushed into the upper mantle. The sinking of the earth's denser, cooler continental crust into its mantle is actually aided by the relatively low density in the upper mantle and also generates the mantle plumes as it sinks towards the earth's core where it heats up and rises to the surface.

2.4.1 The Structure of Celestial Bodies

All the celestial bodies wandering the universe - such as comets, moons, planets and stars - are composed of the same matter that originally comprised the ultimate-body immediately prior to the last 'Big-Bang'. They therefore all have the same age.

When cold (such as galactic force-centres) these celestial bodies consist of the same matter that was created during the previous universal period and held in the ultimate-body. But the celestial bodies that orbit galactic force-centres and have collected satellites of their own, will generate internal heat. These are the stars.

Due to the combined mass of their galactic force-centre and secondary satellites (planets), galactic satellites will eventually generate enough internal heat to create neutrons and heat through fission, the by-product of which is hydrogen (H), at which point they become stars. The core of a star is mostly iron (the commonest and largest stable element), its outer mantle is a store of neutrons and its outer surface comprises hydrogen atoms in the form of proton-electron pairs, which is the reason they can emit electro-magnetic energy.

Fusion only occurs in the ultimate body (and galactic force-centres) where pressure is sufficient to unite proton-electron pairs but produces no heat (energy).

Neutrons are the packets of stored energy that will be contained within the ultimate body at the end of the current universal period and eventually provide the explosive energy for the next 'Big-Bang'.

The *largest* secondary satellites (planets) normally collect sufficient satellite populations (moons) of their own to cause their outer crusts to melt or *'skin'*. These planets will become gas planets.

The term 'skin' means an active [hot] surface covered by skin that forms due to the cooling effect of atmospheric gases.

All *active* celestial bodies (those generating internal heat) will comprise matter of greatest density at their core gradually reducing with radial distance from it. Their structure may be defined by their radial modifier (Δ; refer to Chapter 2.3.1)

2.5 The Atom

Apart from Max Planck's work in the 1920s, everything needed to understand the atom completely and thoroughly was available before 1900. Yet an appreciation of its qualities continues to elude us.

Contrary to all you've been taught about the atom, it is an incredibly simple and elegant design; a brilliant piece of engineering.

The entire universe comprises *only* two particles; the electron and the proton. Both of which are packets of electrical and magnetic [charge] energy.

An atom is simply a collection of *identical* proton-electron pairs fused together under pressure. The number of proton-electron pairs in an atom (or element), along with the neutrons it has acquired, defines its character.

A proton-electron pair is an electron orbiting its proton in a circular path, and if an orbiting electron achieves the speed of light (PE = m.c^2), it will combine with its proton to create a neutron. This is where the universe stores its energy for the next 'Big-Bang'.

Electrons collect energy from electro-magnetic radiation, convert it into kinetic energy, and its proton-partner emits the same [kinetic] energy as electro-magnetic energy.

That is all there is to an atom; **it really *is* that simple**.

In the creation of this book, mathematical models and laws (including atomic heat transfer coefficients) for all the electron shells (n=1 to n=46) in all the elements (Z=1 to Z=92) have been successfully created for any temperature using Coulomb's law of electrical attraction, Newton's laws of orbital motion and the laws of thermodynamics, confirming the validity of this atomic model (refer to Chapter 6.5).

2.5.1 Quanta

Quanta are pictorially represented here as solid spherical objects for convenience only. It is not proposed here that they are in any way spherical, solid or a specific size.

The atom comprises only two atomic particles that between them hold it together and naturally attract or repel other atoms. They are:

Electrons: packets of non-polar magnetic charge, and negative-electrical charge of fixed magnitude and perpetual but variable kinetic energy

Protons: packets of non-polar magnetic charge of fixed magnitude, and positive-electrical charge of variable magnitude

Neutrons: protons and electrons combined through high temperature.

Fig 14

The difference between the strength of attractive and repulsive forces due to magnetic and electrical charges is the **coupling ratio** (φ); 4.407E-40

Gas (a definition)

Similar electrical charges in protons, including proton-electron pairs will prevent them from ever unifying, irrespective of the forces involved.

Therefore, hydrogen atoms (H & H⁺) can neither solidify nor liquefy. It is possible, however, after acquiring an orbiting electron and a neutron. Therefore, deuterium and tritium (D & T) can exist as a liquid, and even as a solid, but hydrogen (H) cannot.

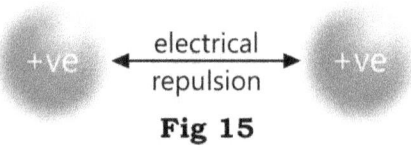

Fig 15

Refer to Chapter 2.5.10 for explanation of *'the state of matter'*.

2.5.1.1 The Electron

The electron is a packet of constant magnetic (m_e) and electrical (e) charges that *must* move.

When in orbit about a proton, the energy an electron collects from its surrounding electro-magnetic radiation is converted to kinetic energy and simultaneously transferred to its proton via their opposite electrical charges (e).

It is not known whether an electron in free-flight is able to absorb electro-magnetic energy [5], but if it can, it will also have the ability to increase its velocity and will thereafter retain this kinetic energy until it can transfer it; via a proton.

Moreover, given its limiting *orbital* velocity (c) and therefore the limitation for naturally created electro-magnetic energy, *if* electrons can absorb electro-magnetic energy in free-flight, this would *naturally* limit free-flying electrons to an upper velocity of less than 'c'.

That said; mass does not vary with velocity and there is nothing to prevent electrons (or anything else) from travelling faster than 'c' if artificially supplied with additional [electro-magnetic] energy.

2.5.1.2 The Proton

A proton is a packet of constant magnetic charge (m_p) and variable electrical charge (e to e').

Its additional magnetic charge provides it with the capacity to hold an operational electric charge (e') that varies with kinetic energy in its orbiting electron.

It uses this operational charge (e') to convert the kinetic energy from its orbiting electron into electro-magnetic energy.

A proton has no kinetic energy of its own.

It does not collect energy from its surroundings. A lone proton (H^+) cannot generate (or emit) energy (e.g. heat) until it traps an electron. If it is not part of a proton-electron pair, its electrical charge cannot be topped up and it can never be unified with another proton.

A lone proton can, however, be forced inside the electron shells of a neighbouring atom if the neighbouring atom has an exposed neutron, thereby altering the characteristics of the neighbouring atom (changing it to another element).

2.5.1.3 The Neutron

A neutron was a *proton-electron pair* in which the orbiting electron achieved the speed of light (c) and became united with its proton. Its magnetic charge is that of an electron plus a proton (m_e+m_p).

When united, the electro-static charges of the proton and the electron will cancel each other out. A neutron's electrical charge is therefore zero. This union is possible only because the magnetic charges of both particles are non-polar.

Therefore, a neutron possesses the same magnetic charge energy as a proton-electron pair, but no electrical imbalance.

Neutrons can only exist inside an atom. They are created inside the atom from the two proton-electron pairs whose electrons are orbiting in the innermost shell. And when ejected, they will revert to their component parts; a pair of protons and a pair of electrons (alpha and beta particles).

It is not yet known what happens [structurally] to the two particles when united as a neutron, or whether a neutron can fully envelop a proton partner; which is unlikely given that this would inhibit electro-magnetic radiation.

A neutron stores *all* the energy that will fuel the next '*Big-Bang*', so it obviously plays an important part in the workings of the universe.

Each neutron holds (stores) the total energy the former proton-electron pair possessed immediately prior to unification, which constitutes;

630,839,312.5 times the energy held by a proton-electron pair when cold (1K) such as in the ultimate-body or outer-space

and

2,309,497.715 times the energy held by a proton-electron pair when warm (1°C) such as on the earth's surface

2.5.2 Proton-Electron Pair

A proton-electron pair is a single proton with a single orbiting electron.

All atoms (elements) comprise collections of proton-electron pairs.

A proton transmits electro-magnetic radiation only whilst it hosts an orbiting electron. It does not emit electro-magnetic energy unless it is part of a proton-electron pair.

An orbiting electron transfers its [kinetic] energy to its proton via its negative electrical charge (e). The proton builds up and holds onto its additional electrical charge (e') via its additional magnetic charge (m_p). The opposite charges in the electron and the proton generate, and transmit, *electron-kinetic* energy in the form of electro-magnetic energy of the same magnitude (refer to Chapter 3.1.3).

Electro-magnetic energy rises and falls with the kinetic energy in the electron, which rises and falls with the energy in the radiation feeding it. In this way, [heat] energy in matter naturally stabilises with the energy in its environment. In outer-space, where there are no proton-electron pairs surrounding it and therefore negligible electro-magnetic energy, matter will naturally fall to ≈0 K

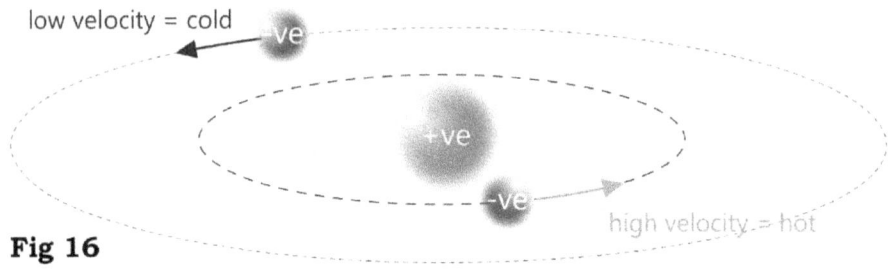

Fig 16

According to Newton's laws of orbital motion, an electron's orbital radius decreases with increasing energy (velocity) (Fig 16). In other words, atomic structural strength increases with temperature.

The electro-magnetic energy generated by proton-electron pairs in an atom is what we understand as heat and light. The combined electro-magnetic energy from all the proton-electron pairs in an atom is its *quantity* of heat. The highest electro-magnetic energy, from the innermost shell, is what we refer to as the atom's temperature.

In addition to electro-magnetic energy, a proton-electron pair also generates a polar magnetic field that travels from the positive (North) face of the orbital plane and around to the negative (South) face of the orbital plane (Fig 17). This is the magnetism that holds neutrons[1] and adjacent atoms together. It is also what you see in bar magnets when the atoms in matter are aligned.

It also generates an electrical field that works contrary to the magnetic field. This energy repels adjacent atoms.

The relative strengths of the electrical and magnetic fields determine whether adjacent atoms exist as a gas or viscous matter (solid/liquid).

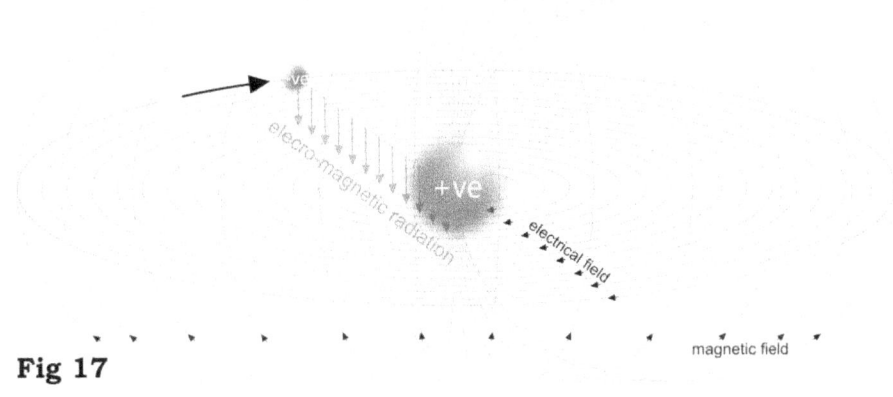

Fig 17

Fig 17 shows the relationship between the electrical and magnetic fields generated by a proton-electron pair.

Electrons ejected from an atom will hold their linear (v) and angular (ω) velocities at the time of ejection in free-flight until affected by impact, gravity and/or electro-magnetic energy. What we see in bubble chambers as post-impact spiral paths is simply the result of impacting electrons that can be visualised as spinning billiard balls obeying Newton's laws of motion.

2.5.3 Electron Shells

Each electron shell contains up to two electrons, both of which are identical. They are held in their shells by a balancing act between their repulsive [same] electrical charges (including the electrons in adjacent shell(s)), and the attractive [opposite] electrical charges with their protons in exactly the same way as Newton described the balance between centrifugal force and gravitational force in an orbiting satellite and its force-centre (refer to Chapter 2.2.8).

Each shell contains up to two electrons diametrically opposite in the same circular orbit. All electrons are identical, except for their kinetic energy if they are not in the same shell.

Shell radii are defined by electron velocity, which is defined by its kinetic energy, which is defined by its heat (surrounding electro-magnetic energy). The highest electron energies are in the innermost shells, gradually reducing with radial distance from the nucleus. Spacing between each shell is identical as it is determined by the repulsion forces between adjacent electrons of identical charge. Key shell radii are defined as follows:

The maximum *possible* shell radius (R_c) is achieved when an electron's velocity (v_c) falls to a level whereby it is expected to leave its orbit and continue in free-flight [3]

Planck's maximum shell radius ($R_o = a_o.(4\pi)^2$) is Planck's maximum orbital radius, the relevance of which is as yet unknown, but happens to be the innermost shell radius of Radon (the largest noble gas) at its gas transition temperature [1]

Planck's mean radius (R_m), is defined by his constant (h) (refer to Chapter 6.11.5) falls between R_o and R_n

The minimum *possible* shell radius (R_n) is achieved when an electron's velocity (c) causes it to combine with its proton and create a neutron (refer to Chapter 6.2.3)

2.5.4 Nucleus

The nucleus of an atom contains protons and neutrons that are held together by <u>magnetic</u> field energy. It is probable that a proton and a neutron may touch (physically) in the nucleus as the neutron will act as a barrier to the positive charge of a neighbouring proton.

Nucleic organisation is dependent upon the isolation of its proton's positive charges by the attached neutrons.

The potential number of proton-electron pair orientations is limited if proton separation is to be ensured. This limitation defines the lattice structures that apply to collections of atoms (refer to Chapter 3.5.3). The structure of any atomic nucleus will be the same as the lattice structure of the elemental matter (at that temperature).

Because of the mirroring of atomic nucleus and elemental lattice structures, the mathematical relationship for elemental lattice structures is the same as that defined by the nucleic factors.

Regardless of the nucleic pattern, structural integrity generally reduces with increasing nucleic size (i.e. as the atomic number increases), making larger atoms generally (but not necessarily) more unstable due to the greater potential for proton-proton and/or neutron-neutron interaction, resulting in radioactivity. Technetium appears to be a special case in which the nucleic arrangement of 43 protons is especially vulnerable to neutron decay.

2.5.5 Electron & Proton Spin

Spin is induced in every particle according to spin theory and varies with the energy absorbed by the orbiting electron.

As a proton has only one satellite (electron), which has no sub-satellites, the electron will always present the same face to its proton throughout its orbit. Therefore, its spin period will be the same as its orbital period:

$\omega = 2\pi/t = v/R$
Where:
v = the curvilinear velocity of the orbiting electron
R = the orbital radius
t = the orbital period

In the event an electron is ejected from its orbit by electrical energy or impact, it will travel in a straight line at its ejected orbital velocity (v) and spin at its latest angular velocity (ω) until recaptured by another proton or impacted by another electron. Because an electron cannot lose its kinetic energy whilst in free-flight and because it will leave its orbit at velocity 'v_c', an electron's minimum velocity must be 'v_c' (refer to Chapter 4.5.1)

Each proton within an atom will spin at a rate commensurate with the orbital velocity of its dedicated electron and in accordance with spin theory.

2.5.6 Isotope

Isotopes are atoms with the same atomic number (Z) but with varying atomic mass because of unequal proton-neutron pairing. Isotope is an alternative way of saying RAM (relative atomic mass).

An atom of iron, with 26 protons (Z=26) and 26 neutrons (N=26) is an isotope of 52. However, in nature, most iron atoms have more than 26 neutrons, each of which is given its own isotope, e.g. 57, 59, etc.

The following rules apply to isotopes:
1) H^+ can never be fused because it only exists as a gas
2) All proton-electron pairs within atoms are Deuterium or Tritium
3) If ψ = N:Z (ratio), then: $1 < \psi < 2$ (see below)

Despite the potential maximum value for ψ = **2**:

*If an atom achieves a 'ψ' value of greater than **1.5** it will eject neutrons as alpha and beta-particles.*

*If an atom achieves a 'ψ' value of greater than **1.6** it will split into smaller atoms ejecting numerous alpha and beta-particles as it does so.*
1.6 *is the limiting number for isotopes.*

Over time, atoms naturally try to achieve ψ = **1**, which is their most stable form. They eventually achieve this by ejecting surplus neutrons as alpha and beta-particles. The rate at which this occurs is referred to as the 'half-life' of the atom. The half-life of any atom appears to be constant, i.e. it never appears to change.

2.5.7 Ion

Ions are atoms with the same atomic number (Z) but possess an electrical charge owing to unequal proton-electron pairing.

Positive ions (atoms that have lost electrons) possess a positive electrical charge. Negative ions (atoms with additional electrons) possess a negative electrical charge. Negative ions are far less common than positive ions.

Only a few atoms exist naturally as negative ions and they are all non-metals$_n$ except for two, which are semi-metals$_s$:

One additional electron (Group VIIA):
Fluorine (9_n), Chlorine (17_n), Bromine (35_n), Iodine (53_n)

Two additional electrons (Group VIA):
Oxygen (8_n), Sulphur (16_n), Selenium (34_n), Tellurium (52_s)

Four additional electrons (Group IVA):
Carbon (6_n), Silicon (14_s).

Any atom can become a positive ion simply by losing one or more of its electrons from impact with free electrons or a strong external positive electrical charge.

Negatively charged ions are a little more difficult to understand. Additional electrons need to be trapped by the positive charge in protons that do not exist in the nucleus: this shouldn't be possible. However, the nucleic structures of the above non-metal atoms probably have at least one exposed proton that is not protected by a neutron and this means that the additional electro-magnetic electrical charge generated in it is available to trap passing free electrons

2.5.8 Radioactivity

A neutron is created within an atom and will be split into its component parts (a proton and an electron) if ejected. I.e. a neutron cannot exist outside an atom.
The minimum neutron-proton ratio is 1.0
The maximum neutron-proton ratio is 1.6, however, technetium (an unnatural atom) becomes unstable with a ratio of 1.3
For clarity, this ratio will be referred to as an atom's *critical ratio*.
Atomic stability is dependent upon nucleic structure.
As an atom's structure approaches its critical ratio, it becomes less stable and ejects unwanted neutrons until a stable ratio is reached. This is called radioactive decay, and the time over which it occurs is referred to as its half-life.

This process releases a great deal of potential energy from the ejected neutrons in the form of electro-magnetic energy, which is generated by the instantaneous creation of a proton-electron pair; the electron of which is orbiting at <'c'. However, in this case, the proton is also travelling at high-speed, freeing it from its partnership.

If an ejected neutron is trapped within the nucleus, this energy will be released as the resultant proton-electron pair becomes incorporated within the atom, changing its atomic number (Z+1).

The potential energy in neutrons released from the atom will be converted into velocity ($v = \sqrt{[2.m_p/PE]}$) that will be close to 7E+06 m/s, giving the proton the capacity to split apart neighbouring neutrons.

A proton will never impact a neighbouring proton because of their similar electrical charges, however, the neutron impact is often sufficient to simultaneously eject the impacted neutron's proton-partner (two protons = alpha-particle). The ejected electrons are beta-particles.

This process will continue unhindered until there are no more available neutrons or the matter has been split apart to such a degree that many ejected neutrons no longer have suitable targets. After which, neutrons ejected from the remaining fragments will impact atoms in the environment.

Atoms close to their critical ratio are subject to a limiting mass condition above which the atomic matter will begin to break apart. This condition is called an atom's critical mass.

As the critical mass is approached, neutron targets in neighbouring atoms are increased resulting in a consequent increase in temperature. Or, if the condition is enforced quickly enough, the ejected atoms will have nowhere to go so the matter will break apart. This is called a chain reaction and is what occurs in an atom bomb.

Whilst our knowledge today allows us only to use such matter in its critical mass, its energy can be released in a controlled manner if processed correctly. I.e. neutron energy can be released from any atomic matter in a controlled and safe manner. Moreover, nuclear waste can provide clean, free energy.

Radioactivity can, and should be regarded as a friend, not an enemy.

2.5.9 How They Work

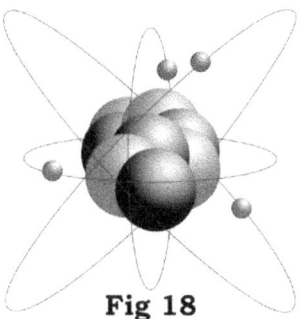

Fig 18

The atom according to Newton and Coulomb is a system that works perfectly, it needs no sub-atomic particles to hold it together, and every electron is identical. Uniqueness and uncertainty are unnecessary. Moreover, apart from Planck's contribution, everything needed to resolve the atom completely and accurately was available before the beginning of the twentieth century.

An atom is a collection of proton-electron pairs. Each orbiting electron in an atom is paired with its own proton (Fig 18).

Their electrical charges hold an electron to its proton, and as the electron has its own perpetual kinetic energy, its orbit *must* be circular.

All elements begin as follows:

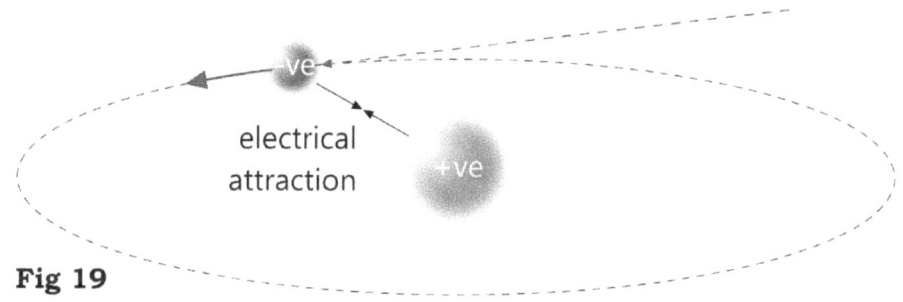

Fig 19

Natural hydrogen comprises mostly lone protons (H⁺) that can trap a passing electron and thus become a proton-electron pair. The majority of hydrogen atoms at the surface of a star, however, are proton-electron pairs as they have been left over from total fissionable dissemination.

Once trapped, the electron will remain in orbit around the proton until one of the following events occur:
1) An adjacent atom provides sufficient excessive electro[-magnetic] charge to cause it to swap orbits, or
2) Another electron impacts it with sufficient energy to knock it out of its orbit, it is then a free electron.
3) Its velocity falls below v_c (refer to Chapter 3.5.4.1) at which its proton can no longer hold onto it

The potential energy (PE = $m_e.v^2$) between the electron and its proton generates an additional static charge in the proton (e'). This additional charge pulls the kinetic energy from the electron and simultaneously emits an equal magnitude (to the electron's kinetic energy) of electro-magnetic energy.

As available electro-magnetic energy decreases, the electron will slow down, reducing the electro-magnetic energy radiated by its proton-electron pairing. If the electron's kinetic energy falls to temperature T_o, the pair could lose its neutron. If its kinetic energy continues to fall to T_c, the electron will eventually leave its orbit and electro-magnetic radiation will cease. The proton's *operational* electrical charge (e') will be lost when it loses its electron.

As electro-magnetic energy available to the electron increases, the energy radiated by the proton will rise accordingly. Eventually, on reaching temperature T_n, the electron's velocity will reach *light-speed* (c). Once this is achieved, the electron will combine with its proton and create a neutron. This pair will no longer emit electro-magnetic radiation, but it will store the energy it had at the moment of unity.

Refer to Chapter 4.5 for a definition of temperatures; T_c, T_o, T_m, & T_n

A proton-electron pair cannot sit alongside another proton-electron pair if neither has collected a neutron, as their similar (positive) electrical charges will prevent this happening.

Electrons and protons possess electrical charges of equal and opposite magnitude when at rest ($N_t \approx 1$). The electrical charge in the electron never changes. That of the proton remains constant and equal to that of the electron when it is not part of a proton-electron pair, but rises with electron kinetic energy as soon as it acquires an orbiting electron.

Lone protons will always consistently repel each other with equal force because of their identical electric charges. This is the most basic form of matter: what we call hydrogen gas (H^+). Lone protons can never accumulate in viscous form because the repulsion force is constant everywhere within the volume they occupy (Fig 15).

An electron's orbit is circular because the electron is providing its own kinetic energy; i.e. it is not generated by the potential energy between it and its proton.

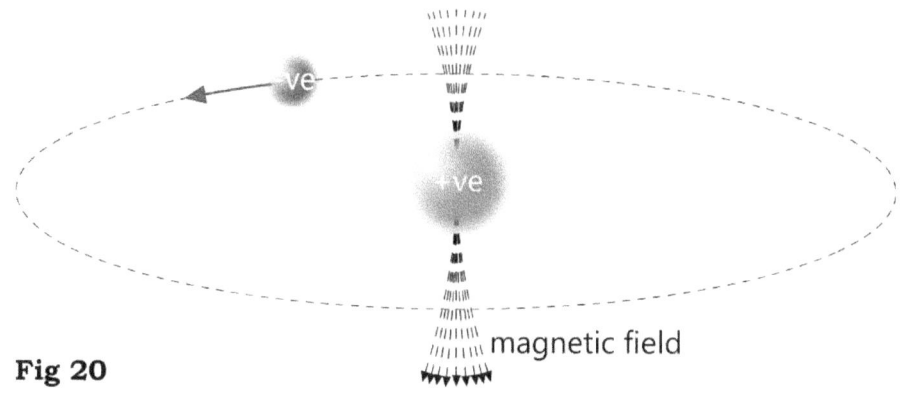

Fig 20

A proton with an orbiting electron naturally generates a magnetic field with a positive (e.g. N) pole at one face of the orbit and a negative (e.g. S) pole at the other (Fig 20), empowering the proton-electron pair to attract a neutron via a magnetic field [1]. This is what we call deuterium (Fig 21).

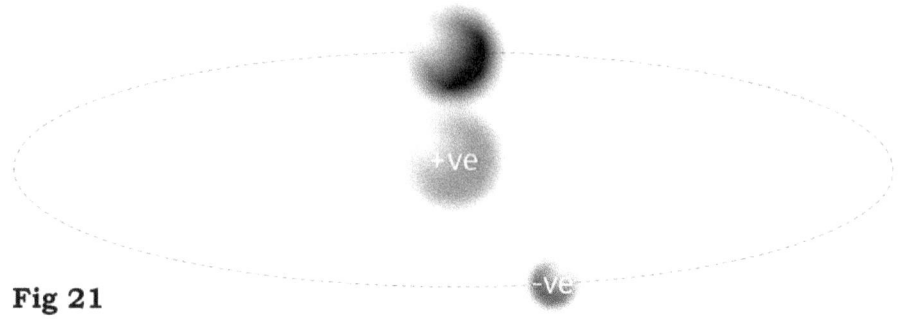

Fig 21

As the kinetic energy in an orbiting electron rises, the potential energy between it and its proton will increase, and will continue to do so as long as all protons within a nucleus are electrically isolated from each other by surrounding neutrons.

2.5.10 The State of Matter

The electrical charge developed in a proton whilst hosting an orbiting electron (e') varies between 'e' at minimum temperature and $\xi_v.e$ at maximum temperature; i.e. immediately before the two particles unite as a neutron. It is this energy that pushes adjacent atoms apart.

The magnetic field energy generated by the proton-electron pair holds adjacent atoms together. This field energy is constant because the magnetic charges generating it are constant (m_e & m_p).

These two competing energies are responsible for the state of matter:

1) If the electron in a proton-electron pair is orbiting slowly (low temperature), the proton's electrical charge (e') will be low, and the electrical repulsion force between adjacent atoms will be less than the attraction force generated by the magnetic field energy. In this case, matter will exist as a viscous substance (solid or liquid); the lower the electron's temperature, the more viscous the matter.

2) If the electron in a proton-electron pair is orbiting quickly (high temperature), the proton's electrical charge (e') will be high, and the electrical repulsion force between adjacent atoms will be greater than the attraction force generated by the magnetic field energy. In this case, matter will exist as a gas; the greater the electron's temperature, the higher the gas pressure.

This is the reason the density of viscous matter remains fairly constant with varying temperature but gas pressure rises and falls proportionally with temperature variation.

Atomic density at low temperatures is considerably lower than matter density. For example;
the density of iron is 7870 kg/m³
whilst the density of its atom:
@ 273.15 K: ρ = 0.083888668 kg/m³
@ 12,412 K: ρ = 7870 kg/m³

At relatively low temperatures, this *'electron-clouding'* allows lone protons (and positive ions) to share *spare* electron charge capacity in neighbouring atoms, but diminishes with increasing temperature. I.e. chemical bonding weakens with increasing temperature.

2.5.11 Fission & Fusion

Fission is the term used to describe the splitting of an atom into two smaller atoms. Fusion is the term used to describe the union of two atoms to create a larger one.

Fusion, the joining of separate proton-electron pairs to create a different element, is accomplished by applying sufficient pressure to force the proton of one proton-electron pair inside the electron shells of another atom. It requires the *input* of energy; it does not generate energy. That is why Hades is cold and 40 years of trials have yet to produce a fusion reactor.

Atoms can only fuse if they are in viscous form; i.e. you cannot fuse gaseous atoms because the electrical field energy, which is greater than the coincident magnetic field energy, will prevent this from occurring (refer to Chapter 2.5.6).

Fission is achieved by releasing the energy stored in a neutron (refer to Chapter 2.5.1.3)

The following formulas are misleading:

i) $D+T \rightarrow {}^4He+\mathbf{n+energy}$

ii) $L_i+\mathbf{n} \rightarrow {}^4He+T\mathbf{+energy}$

iii) $D+L_i \rightarrow 2{}^4He\mathbf{+energy}$

The first law of thermodynamics states that energy cannot be created or lost, it can only be transferred. So, what happens to this energy?

You cannot transfer the neutron from i) to ii) without releasing its energy. That is where the explosion comes from.

"n+energy" is an unreal equation; it should read: "$n \rightarrow H^+ + e^- + energy$"

In order to generate fission, you must first cause a neutron to split into its component parts; a proton and an electron (half of an alpha-particle). Alpha a beta-particles are usually ejected in pairs because there are always two electrons in an element's innermost shell. And because they will always have the same energy at the same time, they will have become neutrons simultaneously. Therefore, they will always be ejected together; two protons and two electrons (same half-life).

The energy released from the split neutron will cause the released proton to be ejected at very high speed (refer to Chapter 2.5.8), impacting a neutron in a neighbouring atom. Whilst it will not impact a neighbouring proton, in practice, a neutron may be ejected together with its proton (an alpha-particle) if the proton is unconstrained, squaring the energy release rate and thereby causing an escalating chain reaction.

Given the rate of ejected protons, and the distances travelled ($\approx 10^{-9}$m) by the alpha-particles, the duration of the chain reaction in a kilogram of matter will be nano-seconds.

An hydrogen bomb does not come from turning deuterium (D) and tritium (T) into helium (H_e), it comes from releasing the energy in their neutrons; i.e. converting D & T into '$nH^+ + ne^- + energy$':

i) $D \rightarrow 2H^+ + 2e +$**energy**

and

ii) $T \rightarrow 3H^+ + 3e +$**energy**

The reason why much more energy is released in the detonation of an hydrogen bomb is because very little of the energy released is lost in splitting the atom, *it is already split*.

The reason the aftermath of an hydrogen bomb is not radioactive is because its by-product is hydrogen. Whereas only a very small percentage of the matter in a uranium or plutonium bomb is converted to non-radioactive matter, the rest of it is scattered as radioactive dust, which is extremely dangerous to anybody coming into contact with it. However, the uranium used to initiate an hydrogen bomb also makes H-bombs radioactive.

2.5.12 The Early Atom

For almost a hundred years, we have been taught that an atom comprises a nucleus of protons and neutrons with electrons orbiting the nucleus in elongated elliptical shells, as proposed by Johannes Rydberg.

His inner shells contained the least number of electrons and the outer shells contain the most. Every electron in the atom must be unique, s, p, d, f, l, m, spin, etc.

It is an enigma (to me), how this level of complexity ever became proposed and/or accepted, because it simply doesn't work.

Fig 22

For example,

1) If Rydberg's shell system is fully analysed, it produces shells of eccentricity equal to 'one'; i.e. a straight line!

2) Elliptical orbits mean variable velocities, which would mean fluctuating electro-magnetic radiation (e.g. colours), which does not occur in reality

3) Electrons in elliptical orbits cannot be equally spaced, and therefore balanced between their similar static electrical charges. This is only possible with circular orbits.

4) Elliptical orbits only work where magnetic potential energy is greater than repulsive electrical energy, which is not the case in an atom.

5) Electrons are not driven by potential energy, as planets are.

6) If the calculated eccentricities are correct, according to elliptical theory; ($b = a.\sqrt{[1 - e^2]}$), the apogee radii reduce with shell number ultimately becoming smaller than the perigee radius of shell-1
(e.g. Shell 6: 2.58899E-11 m)

7) Shell radii calculated according to this theory cannot be reconciled with generally accepted atomic radii for most elements

8) Photographs of atoms show their structure to be spherical, not elliptical

It doesn't matter how the calculations are carried out the Rydberg system cannot be made to work.

Until Quantum theory diverted our attention from reality, we were taught that the atom looked something like the image in Fig 22. The need for an alternative structure came about because Johannes Rydberg's model generated an electron shell eccentricity of 1 (a straight line), which could not be explained or resolved.

Unfortunately, the scientific community unanimously decided that *Quantum Theory* must provide the answer. String theory, sub-atomic particles and the uncertainty principle were subsequently invented to justify it (refer to Chapter 6.3).

This model gave us the fuzzy, non-uniform, statistical, unpredictable, complex shape we frequently see in textbooks today.

In fact, not only are these theories unnecessary, if all matter in the universe actually did obey them, the universe as we know it could not exist. The atom would not work because it could not emit electro-magnetic energy.

After discovering that Newton's laws of orbital motion apply to *all* orbital systems, I decided to see if his theories could be applied to the atom and discovered very quickly that they can.

In fact, the only atom that genuinely works is the Newton-Coulomb atom, and it doesn't need sub-atomic particles to make it work.

So, when reading this book, it is important to remember that; combined with the discoveries of Gilbert, Coulomb, Faraday, Maxwell and Lorentz, Newton's laws of orbital motion also apply to the atom.

2.6 The Universe

The universe is a simple repeating system based upon orbits, neutrons and the coupling ratio.

A galactic orbital system comprises a force-centre, which I refer to as Hades in our Milky Way, together with its satellites (star-systems). All matter resides in galaxies where it is organised by orbital motion. It is all pretty much the same and was created during earlier universal periods and therefore has the same age and properties.

A galactic force-centre comprises matter that was contained within the ultimate-body prior to the '*Big-Bang*'. Its satellites are simply bodies of the same elemental matter orbiting according to the kinetic energy induced by the '*Big-Bang*'.

For example: our sun comprised similar elemental matter as the earth when originally ejected by the '*Big-Bang*', but its internal frictional heat is sufficient to convert its matter to proton-electron pairs. The earth doesn't generate sufficient internal heat to melt its crust as it has just one moon.

After collecting enough *planetary* satellites, the competing orbital kinetic and potential energies will generate sufficient internal [frictional] heat in a galactic satellite to melt its outer crust turning it into a gas satellite. As its planetary satellite population increases still further (from galactic comets), the internal [heat] energy generated will eventually be sufficient to convert proton-electron pairs into neutrons through fission, reducing elements as it does so. It will then have become a star.

The hydrogen gas (proton-electron pairs) generated by a star will migrate to its surface. This surface hydrogen is what we see and gives a star its signature (refer to Chapter 6.5.4). A star is *not made from* hydrogen (H^+) and it does not generate fusion, it *generates* hydrogen (H) through fission. If fusion generated heat, Hades would be hot, but it isn't, it is cold.

As a star converts its matter to hydrogen, it becomes colder. When it has insufficient elemental matter left, it will become a cold mass of neutron-rich matter and light elements surrounded by a large cloud of proton-electron pairs. Neutrons alone cannot generate electro-magnetic energy through internal friction because frictional heat is generated by opposing electrical charges in electrons. Neutrons possess no electrical polarity and therefore cannot generate electrical fields.

Gas planets are simply stars with much less energy. The largest planetary satellites always trap the most galactic comets (moons) and therefore generate sufficient internal frictional heat to melt their outer surface and vaporise the lighter molecules. They are unlikely, however, to generate enough heat to create neutrons due to their smaller force-centre and satellite masses.

If the earth acquired another substantial moon, it is quite possible that the internal (mantle) heat generated would be sufficient to melt its crust. The earth would then become a gas planet. Mars is not a gas planet because it has almost no mantle material ('Δ'; refer to Chapter 2.6.3.5)

As galactic force-centres have no force-centre of their own, they cannot generate internal frictional heat, so they are cold (dark).

There may be many of these [dark] bodies in our galaxy but just because they don't radiate energy, doesn't stop them continuing on their way around their galactic orbits just as they did before. They will possess the same *mass* after their death as they did during their active life.

Galaxies will remain orbital systems even after all their stars are dead and cold. They will only collapse into a single mass when they eventually re-aggregate into another ultimate-body.

No matter how close a satellite is to its force-centre, its centrifugal and gravitational forces *must* balance. Other than collision, nothing in its orbital progress will cause a satellite to be consumed by its force-centre. When adjacent orbital systems collide, one or both of the orbital systems may be destroyed or combined through impact.

Just as in solar systems, galaxies have their own comets, but they are likely to be much larger than the solar variety. The passage of galactic comets through a solar system could provide additional planets or moons, or disrupt solar satellites through impact and/or tip them upside down, reversing their magnetic polarity in the process.

Universal expansion following the '*Big-Bang*' is gradually being slowed down by inter-galactic magnetism. Eventually, all universal matter will stop travelling outwards and begin to travel back together, where it will eventually reunite. As this mass accretes, core-pressure will create new heavy elements through fusion. Eventually, after *all* the matter in the universe has recreated an ultimate-body, 'bang' it will start all over again.

2.6.1 Life, the universe & everything

In the beginning there was an ultimate-body of >2.8E+75 Quanta, most of which was in the form of matter created in the planets, stars and galactic force-centres during previous universal periods. Elements are created through fusion in the core of this body.

As universal matter accretes into an ultimate-body, pressure at its core will become sufficient to compromise the integrity of its innermost neutrons due to the coupling ratio (φ). The resultant chain-reaction (*Big-Bang*) released the energy held within sufficient neutrons to eject almost all of the Quanta in clumps of varying size and density that become the stars, planets, comets and galactic force-centres. The number of Quanta in the ultimate-body is the same as that in the universe today.

During each universal period, lighter elements and neutrons are created through fission in the stars. The energy (required for the next *Big-Bang*) is stored in these neutrons.

Localised magnetic attraction collected the Quanta into orbiting groups that eventually became galaxies. The largest masses ejected from the ultimate-body became galactic force-centres.

The potential energy of each galactic force-centre and the slightly different kinetic energies in the galactic bodies created the galactic orbits.

All galactic satellites (stars) and sub-satellites (planets) are collectively responsible for generating all the electro-magnetic energy in the universe through internal friction (spin).

All other universal bodies, such as galactic force-centres, moons and stars with no planets will be dark because they generate no internal frictional heat.

This means that every star (that we can see in the night-sky), must by definition be hosting a substantial planetary satellite system.

Whilst most galaxies appear to be moving away from each other, this is due to the relative 3-D nature of post '*Big-Bang*' galactic travel (Hubble's law). Galaxies are actually slowing down with time, albeit gradually, owing to universal magnetism.

Once a galaxy stops moving away from its origin, magnetism takes over. All galaxies will eventually be attracted to each other until reunited as an ultimate-body and 'Bang', it starts all over again.

The above sequence complies with the three laws of thermodynamics and can repeat itself eternally with no outside help.

The universe is in-fact one enormous energy generator, and this is how it works:

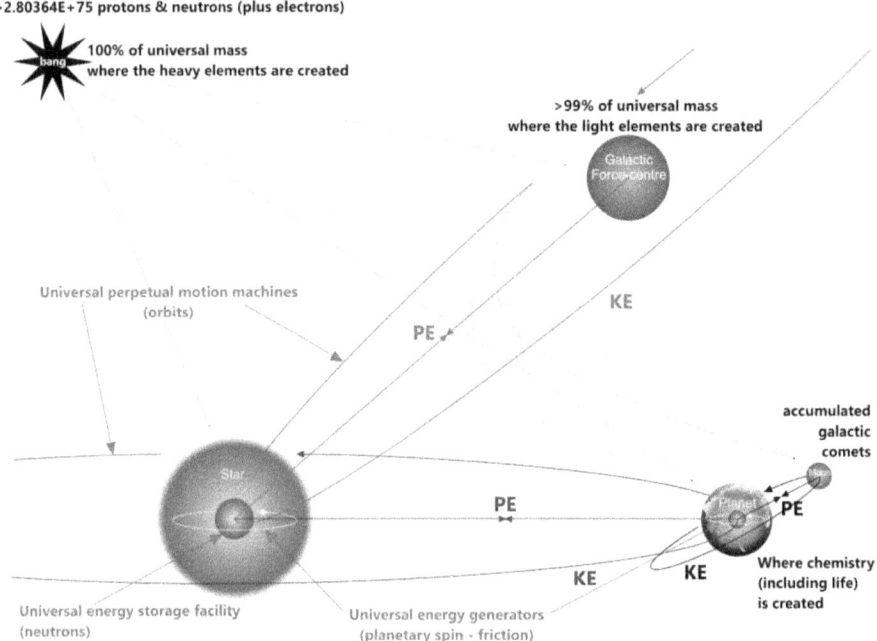

2.6.2 The Milky Way

A detailed analysis of the solar system reveals the correct amount of matter together with all the velocities and orbital shapes with no inconsistencies. This is how we know that Isaac Newton's theories are correct. Contrary to popular belief, the same can also be said of our Milky Way galaxy.

According to a scientist in the 1930s, Newton's laws predicted that the total mass in our galaxy (the Milky Way) could not be accounted for from observations and suggested that the entire Milky Way must be full of dark matter (which comprised sub-atomic particles) to prevent the spiral arms of the galaxy being ejected under centrifugal force.

In fact, this claim has now escalated to the extent that it has been embraced by 99% of the world's astrophysicists who believe that 85% of the entire universe comprises dark matter in the form of sub-atomic particles, and a great deal of time, money and effort has been spent looking for it (refer to Chapter 2.7.1).

It is difficult to understand this situation given that assuming NASA's data for our sun's perigee distance, its orbital period and the estimated [number of] solar masses in the Milky Way are all correct, the sun's orbit within the Milky Way can be fully resolved using Newton's laws of motion and planetary spin theory. Even if NASA's data is incorrect, the same calculation can be carried out using different input data revealing slightly different results but no less successfully. There simply isn't any need for dark matter.

2.6.2.1 Hades

Hades is a name I have adopted for the Milky Way's force-centre for easier reference.

We know it exists because it is a fundamental law of nature that every orbital system *must* have a force-centre (refer to Chapter 2.2.1).

So why can't we see it?

Hades has no force-centre of its own, so it cannot generate heat energy (through internal friction). It is therefore cold; it emits almost no electro-magnetic radiation.

Moreover, it has a diameter of ≈3.5E+12m (refer to Chapter 3.3.6),
whilst our Milky Way has a major axis of ≈1E+21m
the ratio being; 3.5E-09

So, it would be like looking for a dark atom in the centre of a 1m diameter disc of black-iron.
That's why you can't see it.

But it *is* there, and it is the primary source of the heat energy generated in all its orbiting stars.

2.6.3 Our Solar System

A solar system is a collection of satellites (planets and comets) orbiting a star. Whilst the star is active, it will continue to generate fission and thereby radiate heat and light (electro-magnetic energy) to its satellites. The nearest orbiting satellites will receive the most heat and light.

A solar system comprises orbiting bodies that usually include; planets, moons, comets and meteorites. Those that orbit a star are generally planets and comets. Meteorites may also be contained within lunar orbits, along with moons.

All the bodies in our solar system originally comprised the same matter. The difference between the planets is governed by:
1) Their proximity to a significant heat source
And
2) The immensity of their force-centre and satellites (internal spin energy)

Out-of-plane alignment and high orbital eccentricity in a comet (satellite) are both strong clues that a body has been acquired from outside the solar system (galactic comets).

2.6.3.1 Our Sun

Δ = 0.318284697814735   ρ = 1409.782932 kg/m³

As with all stars, our sun originally comprised the same elemental matter as all the other celestial bodies in the universe. However, its massive force-centre (Hades) and significant satellite population (our solar system) is sufficient to generate the internal [frictional] heat required to convert elemental proton-electron pairs into neutrons, the by-product of which is hydrogen (H; proton-electron pairs).

It is currently claimed that our sun is creating elements from hydrogen (H⁺) through fusion and that it is [apparently] growing in size with age. The problem with this scenario is that fusion *increases density* and therefore *reduces size*. And why is Hades cold if fusion generates heat, as Hades is far more likely to generate fusion than our sun.

Its hot hydrogen blanket is a star's atmosphere, which generates the electro-magnetic energy we receive as light and heat (along with other forms of electro-magnetic radiation). The only reason we can see it is because the hydrogen at its surface comprises proton-electron pairs, which *are* capable of emitting electro-magnetic radiation (refer to Chapter 6.5.4). Our sun's atmosphere would be invisible (and cold) if its surface was natural hydrogen (H⁺; lone protons).

The neutrons created in all stars, including our sun, provide the energy (refer to Chapter 2.5.1.3) that will ultimately fuel the next '*Big-Bang*'.

2.6.3.2 Mercury

$\Delta = 0.81286219642311 \qquad \rho = 5427.012135 \text{ kg/m}^3$

Mercury's 'Δ' value and its density show it to be an iron-rich planet with an internal structure much more evenly distributed than is the case for Venus or the earth.

Because it has no moon to generate internal friction it cannot have a particularly active core and is therefore unable to generate any internal friction or volcanic activity, hence no surface gases (atmosphere). This is also the reason why the surface of the planet's '*far-side*' is so cold, despite being so close to the sun; it has no internally generated heat.

Mercury's matter is similar to when it was ejected during the '*Big-Bang*'.

Mercury is therefore a solid piece of iron-rich matter with a cold core. There will be little or no volcanic activity on Mercury as it has no tectonic plates and negligible internal heat.

There is little to distinguish it from, say, Pluto other than its composition and the temperature of half its surface, and that Mercury has acquired no moons of its own because it is too close to the sun. Any passing galactic comets will have been trapped by the sun's greater mass and proximity.

2.6.3.3 Venus

Δ = 0.681231909980155   **ρ = 5242.664311 kg/m³**

Venus's 'Δ' value and its density show it to be an iron-rich planet with an internal structure more evenly distributed than is the case for our earth but less than that within Mercury. Its origin is similar to that of Mercury (the ultimate-body).

It spins the opposite way to the other planets because: a) it has no moon(s) to drive it in the other direction, and; b) owing to its size, the sun's angular kinetic energy dominates Venus's spin.

Its much greater mass (than Mercury), however, will mean that whilst it has no moon to generate a significant level of internal friction, its mass is sufficient to induce some internal friction (heat) …

$E_2 = E_1 - E_0$ ($E_3 = 0$)

… giving it the capacity to generate some (minimal) volcanic activity, and therefore surface gases (atmosphere).

Venus's internal matter is therefore more likely to gravitate towards its core than within Mercury. This also means that whilst there will be insufficient heat to generate tectonic plates (Venus's crust is too thick), there should be sufficient to generate sporadic (and random) volcanic activity, which together with its proximity to the sun, will generate more than enough heat to vaporise its atmospheric water.

The reason Venus's surface material is so flat is because its volcanic activity is far less aggressive than on the earth due to its significantly lower internal heat and much thicker crust.

Venus is, and always has been, too close to the sun to allow water to exist on its surface in liquid form. The mass of water vapour maintains the surface temperature of Venus.

2.6.3.4 The Earth

Δ = 0.3342776982996   **ρ = 5506.351327 kg/m³**

Unlike Venus, the earth has a substantial satellite driving its angular motion, contrary to which, the sun is trying to drive the earth's core in the opposite direction and slow it down. Therefore, the earth's mantle and its core are revolving at different rates, generating internal friction and, as a result, internal heat.

This relative rotation (7E-05 ^{c}/s – refer to Chapter 3.3.2) is sufficient to generate the internal heat through friction that drives the mantle plumes, and in turn drives the tectonic plates. The earth gets its earthquakes, volcanoes, weather and its carbon-cycle from this internal activity. The moon's tilted orbital plane gives the earth its seasons. Along with the earth's proximity to its sun, all this internal and surface activity provides an ideal environment for life to flourish.

Spin and core-pressure analysis has shown us that the density of the earth's mantle material just below its crust is not much more than that of liquid water, allowing heavier material, such as mountain roots to fall from the crust.

Referring back to earlier attempts to age the earth via heat loss:

Kelvin's assessment of the age of the earth through heat loss may not have been wide of the mark if the earth's internal heat is indeed left over from its birth. But all this tells us; is that if his hypothesis was correct, the earth would have lost all of its internal heat within the first 20 to 100 million years of its life, which is clearly not the case.

Rutherford's subsequent claim that additional heat sources would have increased the earth's perceived age was, however, wide of the mark, as there is no way the heat generated by the earth's radioactive substances could account for the heat it currently possesses.

The magnitude of the earth's internal heat means that it must be constantly generated, and its only source can be from friction between its inner core and its mantle matter. This heat is the source of the earth's mantle plumes. It is greatest at the core-mantle interface and rises to cool below the earth's crust.

Less than 0.04% of earth's atmosphere is generated by its mantle activity, all the rest (N_2, O_2 & Ar) has been created by its surface life and potassium decay. Less than 40% of the earth's atmospheric heat is generated by the sun's radiated energy. The rest (>60%) is generated by its mantle heat.

The early atmosphere must have been very thin indeed (almost non-existent): mainly sulphurous chemicals, hydrogen and carbon dioxide, but with only a tiny fraction of today's density and volume (<0.039%). Therefore, meteorites would have been considerably colder at the time of impact than would be the case today. Only friction generates heat, not compression and rupture. Given this and the following:

a) The earth's internal heat is generated by internal friction.
b) Before acquiring a moon, the earth was cold.
c) The earth's surface water was liquid well before 4bn years ago.
d) Impacting masses create very little heat if they don't have to pass through a heavy atmosphere (no friction).

The earliest period of earth's existence must have been quite cool. It will have had no moon and therefore no tectonic plates or volcanoes. Its first volcanoes, after acquiring its moon, would have been distributed at random. More than 3.8bn years ago there were pebble beaches on the surface of the earth (The Isua Greenstone Belt, Greenland) indicating that it must have had active surface water at least 4bn years ago. It is highly likely therefore, that the images we see of a hot earth during creation are incorrect as there was very little to generate the heat and there appears to be no supporting geological evidence. Moreover, the earth was never born, like all the other celestial bodies in the universe; it was ejected from the ultimate-body 13.6bn years ago.

The meteorites we collect that have been aged at 4.66bn years old are nothing more than left-over rubble from the destruction of the planet that once orbited at the asteroid belt.

2.6.3.5 Mars

Δ = 0.0023170868178197   **ρ = 3934.080869 kg/m³**

It is assumed that the information sent back from the various planetary missions to Mars have confirmed its mass.

If Mars is also an iron-rich planet, its density and 'Δ' value would tend to indicate that the planet must be hollow.

It is probable that the energy in Mars' largest moon (Phobos) was sufficient to blast its iron-rich mantle material onto its surface, leaving its internals full of voids, which would explain its apparent density and its 'Δ' value. All of its surface water would no doubt have migrated into its these voids and is occasionally released only under heavy meteorite impact.

If correct, its exceptionally low 'Δ' value, low density, red colour and large rapid primary moon (Phobos) suggest that Mars may well be an iron planet that had a short ultra-active life because of a fast-spinning mantle that would have been responsible for creating such enormous volcanoes for such a small planet.

During this early activity, sufficient internal heat would have been generated to maintain Mars's surface water in liquid form despite its distance from the sun and may also have been able to generate life earlier than the earth. Mars's red surface colour (rust) may well indicate that it had accommodated oxygen-emitting plant life along with liquid surface water before its mantle was blown out.

Apart from the earth, Mars is by far the most interesting planet in our solar system, because *if* the following are true:
1) it is a hollow iron planet,
2) it had liquid water on its surface,
3) it has hosted oxygen-emitting plant-life,
4) its water has found its way into the planet's interior,
… it is possible that the kinetic energy in Phobos is keeping Mars's internal water liquid, which would mean that:
a) it may contain an atmosphere (of some description) internally,
b) it may contain life internally, and
c) it may be possible to access its internal voids.

Could it be that there is an atmosphere and water inside Mars, being kept liquid by an over-active moon?

2.6.3.6 The Asteroid Belt

$\Delta \approx 1$ (est.) $\rho = 2090$ kg/m³

It appears that there was once a planet at position '5' in our solar system and that it was impacted by a galactic comet, and is probably the source of many of our meteorites today and in the past. Perhaps this impact created the comet that hit the earth 63mn years ago; i.e. the asteroid belt was created 63mn years ago.

One of the reasons we know that our sun and its planets did not accrete from rocky particles is that the Asteroid belt remains a loose collection of rocks. I.e. planets do not accrete through gravity.

Because the moons in our solar system are trapped from galactic travel, it would appear from the Asteroid incident, that galactic comets can be quite substantial. However, whilst our own comets tend to orbit in cycles of hundreds of years, galactic comets will orbit in millions of years. But keep your eyes open!

2.6.3.7 Jupiter

Δ = 0.0227806696137   **ρ = 1326.216812 kg/m³**

Jupiter comprises the same matter and has the same age as every other planet, star, comet, etc., in the universe, i.e. that of the current universal period.

Jupiter is a gas planet simply because its *mass* and orbital distance have together enabled it to trap many sizeable moons; Jupiter's largest moon (Ganymede) is approximately twice the mass of the earth's moon. Together with the mass of its force-centre, its satellite population (>50) is sufficient to generate the heat necessary to melt its crust, but not sufficient to generate neutrons (fission). This is the reason for its apparently low density. Its average body density should be calculated on the basis of an outer diameter that excludes its gas cloud and is likely to be similar to that of the solar system's inner planets.

Whilst its surface may be molten, its heavy surface gases will probably extract sufficient heat to form a protective skin over its surface, but it will be relatively thin.

Jupiter's weather is created by its moons orbiting in both directions (prograde and retrograde) that together generate sufficient competing kinetic energies in the planet's surface gases to account for its violent weather. The Red-Spot rotates because of its two adjacent gaseous layers rotating in opposite directions, which is caused by the opposing orbital directions of its moons.

If Jupiter's density is similar to that of the inner planets (5392 kg/m³); its body diameter should be 87605252 m (6.875 times that of the earth), giving it a gas-cloud cover thickness of 26107374 m

Using this information, the body of Jupiter is spinning at 1.9344×10^{-5} radians per second and its properties are:

Δ = 0.33   **ρ = 5392 kg/m³**

2.6.3.8 Saturn

Δ = 0.0140600109265   **ρ = 687.1230137 kg/m³**

Just as for all the universal bodies, Saturn comprises the same matter and is of the same age as every other planet, star, comet, etc., i.e. that of the current universal period.

Whilst Saturn is only 30% of Jupiter's mass, its largest moon (Titan) is almost 92% as massive as Jupiter's largest moon, therefore Saturn will be much more active [internally] than Jupiter.

This is no doubt the reason for its extremely low density. A significant percentage of its body matter has been converted to gas and is likely to generate more active weather than Jupiter.

Saturn's rings are most probably the remains of a satellite that was pulled apart by the huge gravitational forces induced by orbiting so close to such a large planet. If the satellite was an ice moon, it would not have required much kinetic and potential energy to pull it apart.

Saturn and Jupiter have collected the most moons because they are the most massive planetary satellites in our solar system.

If Saturn's density is similar to that of the inner planets (5392 kg/m³); its body diameter should be 58607472 m (4.6 times that of the earth), giving it a gas-cloud cover thickness of 28926264 m

Using this information, the body of Saturn is spinning at 1.6379E-05 radians per second and its properties are:

Δ = 0.28   **ρ = 5392 kg/m³**

2.6.3.9 Uranus

Δ = 0.0249376193237   **ρ = 1270.415139 kg/m³**

Uranus is another 'gas' planet just like Jupiter and Saturn for exactly the same reasons, but being so much further away from its force-centre's radiated heat, much of Uranus's surface gases will remain in liquid form, hence its relatively (to Saturn) high density.

Whilst Uranus has at least 30 moons, all but the smallest, which are also the furthest away, orbit in the same direction. It is not therefore expected that the same level of disruption will be seen on the surface of Uranus as that seen on Jupiter and Saturn.

If Uranus's density is similar to that of the inner planets (5392 kg/m³); its body diameter should be 31328902 m (2.46 times that of the earth), giving it a gas-cloud cover thickness of 9697549 m

Using this information, the body of Uranus is spinning at -1.5186E-05 radians per second and its properties are:

Δ = 0.27   **ρ = 5392 kg/m³**

2.6.3.10 Neptune

Δ = 0.06523792740988   **ρ = 1637.934377 kg/m³**

Neptune is also a gas planet, but so far from its sun that much of its surface gases will remain in liquid and even solid (ice) form, hence its greater (than Uranus) density. That said; it still has a satellite population sufficient to melt its crust.

Whilst Neptune has only 15 or so moons, its largest moon (Triton) orbits in the opposite direction to most of the others. It is expected therefore that Neptune's surface matter will suffer significant levels of disruption.

If Neptune's density is similar to that of the inner planets (5392 kg/m³); its body diameter should be 33103096 m (2.6 times that of the earth), giving it a gas-cloud cover thickness of 8070452 m

Using this information, the body of Uranus is spinning at 3.7918E-05 radians per second and its properties are:

Δ = 0.28   **ρ = 5392 kg/m³**

2.6.3.11 Pluto

Δ = 8.64241984998   **ρ = 1859.960193 kg/m³**

Pluto is a small ice planet in which all of its matter exists in solid form, hence its relatively high density.

Having acquired a substantial sub-satellite population, the largest of which (Charon) is more than 10% as massive as the planet itself, the competing kinetic and potential energies are pulling Pluto into a localized orbit, and is the reason it has a 'Δ' value greater than 1 (refer to Chapter 4.3.1).

Due to its localised orbit, the sun's potential energy is not acting at the planet's centre. I.e. there is no conflicting core-mantle spin to generate internal heat (energy), otherwise, Pluto would actually be a gas-planet.

Pluto is more entitled to be called a planet than either Mercury or Venus, because whilst all three are solid lumps of matter, at least Pluto has managed to acquire some moons; at least five in-fact. Moreover, if it had attracted less massive sub-satellites, it would have an active core, just as all the other planets in our solar system; except Mercury and Venus.

Whilst there are planets outside Pluto's orbit; MakeMake, Haumea, Eris, etc., as we know very little about them, I have not addressed them here.

2.6.3.12 Moons & Comets

A lunar system is a collection of satellites (moons and perhaps meteorites) orbiting their force-centre(s); planet(s).

All the lunar orbital planes in our solar system seem to indicate that most if not all its moons are foreigners, because:

1) If they were accreted from elemental matter, they would be orbiting in the same plane as the planets.
2) They would comprise the same matter as their planetary force-centre
3) If comets arrive from outside the solar system (which is highly likely) there is no reason to disbelieve the same origin for our moons
4) Galactic orbital systems must surely have their own comets, which is probably their origin.

The above is a perfectly reasonable supposition given that, apart from Jupiter, there is a significant tilt in all the lunar orbital planes in our solar system indicating that none of the moons were orbiting during the early solar system. The gravitational pull between Jupiter and our sun, however, was large enough to overcome a tilt that would otherwise exist, thereby allowing Jupiter to align its spin axis with the sun (refer to Chapter 2.2.9).

We may be fairly sure therefore, that apart from asteroid belt residue, many (if not most) of the moons in our solar system were trapped whilst passing through, i.e. they were originally galactic comets.

If moons *are* pre-formed bodies trapped by planets, i.e. not accreted, it is probable that planets orbiting closest to their star are likely to be satellite-free. The star's mass and proximity would normally be sufficient to ensnare or deflect any celestial body that would otherwise have been trapped by the innermost planets, and would be the reason why, for example, Venus and Mercury are the only satellite-free planets in our own solar system, and why the earth has only one.

It is difficult to assess the 'Δ' value of a satellite that doesn't spin, but an estimate may be made based upon its average and surface densities.

If a planet's mantle is hot and therefore of low viscosity, its destruction through impact is likely to result in clumps of ejected matter of varying density, that gravitational energy can reform into spheres (the lowest form of energy). This may be the reason for moons of varying density.

If this was the case with the planet that orbited at the asteroid belt, that planet must have had a substantial lunar satellite system – as is the case for all our solar system planets apart from Mercury and Venus –, which means it will have been able to generate volcanic activity and thereby create rocks. This, in turn, makes it highly likely that this ex-planet is the source of the comets we collect here on earth, which are made from rock.

Our Moon

$\Delta = 0.554903434 \qquad \rho = 3343.599878 \text{ kg/m}^3$

As our moon comprises very different matter (3343.6 kg/m³) from the iron planets and its 'Δ' value shows that it is unlikely to be hollow, it must have come from somewhere else in the galaxy or solar system and therefore must have been acquired after the earth began to orbit its sun.

However, given the estimated density for the largest mass in the asteroid belt (Ceres) is 2090 kg/m³, it is a fairly safe bet that our moon originated from outside our solar system, especially given the inclination of its orbital plane.

Phobos

$\Delta = 0.275895222 \qquad \rho = 1827.4186 \text{ kg/m}^3$

The same would normally go for Phobos, but its density is very similar to that of Ceres, so Phobos may be left over from the impact that created the asteroid belt, especially given their orbital proximity.

Deimos

$\Delta = 0.01434654 \qquad \rho = 1471 \text{ kg/m}^3$

Deimos is probably hollow and also appears to comprise matter different to that which would be expected in its part of the solar system. But if the Asteroid planet was capable of creating rock, it is possible that Deimos was part of the Asteroid's upper mantle material, or perhaps one of its moons.

2.7 Fact & Fiction

There are a number of myths prevalent in the scientific community, not least the Theory of Relativity and Quantum Theory (refer to Chapters 6.2 to 6.4). These obscure theories have ultimately led to mystical features such as anti-matter, sub-atomic particles, dark matter, Black Holes (which are actually black bodies) and singularities.

2.7.1 Sub-Atomic Particles (fiction)

These sub-atomic particles along with electron uniqueness; s, p, d, f, l, m, spin, etc. were invented to hold Bohr's atom together and make it work. Whilst sub-atomic particles, such as gluons, fermions, bosons, positrons, etc. may or may not exist, none of them are required to make the universe – as we know it – work; they are *unnecessary*. And if there is one thing certain about nature, it doesn't waste energy on things it doesn't need.

The real atom (refer to Chapter 2.5) doesn't need them, energy transfer (electro-magnetic radiation and fields) doesn't need them, galactic structural integrity doesn't need them, so why would they exist?

For example; there is no need to split a proton into quarks (1-Down and 2-Up). An electrical charge will *always* have a magnetic counterpart (a magnetic charge); it is nature's fundamental design. It is a perfectly understandable, workable and reliable particle; it possesses everything it needs. There is nothing to go wrong and it is 100% efficient.
Why would you make it complicated and therefore unreliable by giving it component parts it doesn't need, that could go wrong and wastes energy?

Not only are these particles unnecessary, there is no physical evidence to show they exist. E.g. *Bubble-Chamber* diagrams reveal interactions that can be easily explained as collisions between electrons in free-flight.

The Hadron Collider (CERN) was built, at enormous cost, to identify and define all these sub-atomic particles, which is a shame because they most probably don't exist, or at least not necessary to make the real atom work.

Anti-matter is also a fallacy. There is absolutely no argument that can explain or justify its need in our universal system.

The real atom works perfectly because it is a brilliantly elegant, simple, reliable orbital system of just two *perfect* particles; the electron and the proton. Everything we see and feel around us can be explained by them.

The atom described by Quantum Theory, which is not based upon orbital systems, cannot work because it cannot transfer energy.

Together with the *uncertainty* principle, the invention of these particles was needed to justify an atomic system that cannot work without them. Moreover, they remain unproven and undiscovered to this day (refer to Chapter 6.3).

2.7.2 Black Holes (fiction)

I'm afraid there is no such thing as a singularity or an event horizon.

The reason stars emit intense heat and light (electro-magnetic energy) whilst black-bodies do not, is because stars have a force-centre and a substantial satellite population. The heat energy generated by the resultant internal friction unites protons and electrons as neutrons. Smaller elements are generated in galactic force-centres through fusion, but this is not a source of heat.

Fission releases the electro-magnetic energy we see and feel as light and heat from an active star that is also creating hydrogen and neutrons. As more and more of a star's matter is converted to hydrogen, it will grow in *physical* size (not in mass). When there is insufficient viscous matter to generate the frictional heat at its core the star will become inactive (cold). This black-body will eventually become a hydrogen (proton-electron pair) cloud surrounding a neutron-rich core.

All black bodies, which can be any size, are simply cold. They do not have force-centres and/or satellites sufficient to generate internal friction.

Galactic force-centres, which are invariably black-bodies, spin very slowly indeed. Hades, for example, is spinning at about 2E-07 radians per second.

If we were to define a *photon* as an electron travelling at the speed of light and a black-hole as a black body large enough to trap *photons*, then;
$PE_{bh} \geq KE_e$

Where; PE_{bh} is the gravitational energy at the surface of a black hole and KE_e is the kinetic energy in a photon:
$PE_{bh} = G.m_1.m_2/R = 6.1350483563E\text{-}12$ J
$KE_e = \frac{1}{2}.m_2.c^2 = 4.093555584E\text{-}14$ J
Therefore, Hades has more than enough gravitational energy to trap a *photon*.

The minimum size for black-hole comprising mostly iron, may be calculated as follows:
$E = G.m_1.m_2/R = \frac{1}{2}.m_2.c^2$
Given that; $R = \sqrt[3]{3.m_1 / 4.\pi.\rho}$
$m_1 = \sqrt{[3.c^6 / 32\pi.\rho.G^3]}$ = **9.623785516E+37 kg**

Therefore, Hades is larger than the minimum sized [hypothetical] black-hole (refer to Chapter 3.3.6).

In order to increase the density of a black-body above that of iron, gravitational (magnetic) pressure must exceed the electrical resistance by at least $1/\varphi$. The minimum mass for what is referred to as a black hole (9.6238E+37 kg) is insufficient to achieve this. The density of the smallest black hole must be; 7870 kg/m³, have a radius of ≈1.7496567E+12 m and a polar moment of inertia of ≈2.1622E+65 kg.m², because:

The repulsion force between adjacent iron atoms may be calculated using Coulomb's law:
$F = k.Q^2 / R^2$
Where: Coulomb's constant; k = 8.98755184732667E+09 kg.m³ / C².s²
Elementary charge unit; Q = 1.60217648753E-19 C
Proton spacing; R = 2.8E-10 m
F_e = 2.9427E-09 N between any two atomic nuclei.

Iron has 8 of these atoms in any lattice structure, so:
F = F x 8 = 2.35416E-08 N
The surface area at R will be: A = 4.π.R² = 3.284E-19 m²
Positive pressure at R: p_e = F_e /A = 7.1686E+10 N/m²
Gravitational pressure necessary to increase core density must be:
p_e/φ = 7.1686E+10 / 4.40742E-40 = 1.62648E+50 N/m²
This pressure occurs within the smallest black-body at a radius of 1.03617E-19m, which is 2.7E+09 times smaller than the proton spacing (R) above and therefore means that the gravitational pressure at the centre of the minimum black hole is insufficient to compress iron atoms and thereby alter its density.

However, as light is not photons (refer to Chapter 1.1.1) and black bodies are simply cold and can be any size, they cannot be defined by their ability to trap electrons. The above calculations are simply an hypothetical mathematical exercise.

2.7.3 Big-Bang (fact)

The energy released within the ultimate-body has popularly become known as the *'Big-Bang'*.

The *'Big-Bang'* originated from the energy released by compromising the inner-most core neutrons (the stored energy within each being 4.1E-14 J) within the ultimate-body of >2.80364E+75 Quanta. The core of this body is where all the heaviest universal elements are created. It is only inside the core of a body of this mass that there is sufficient pressure to generate the necessary fusion.

The pressure at its core was sufficient to overcome the coupling ratio (φ) splitting the innermost neutrons into their component parts and releasing their stored energy in the form of alpha particles. The resulting chain reaction released >4.5E+56 J of energy that caused the ultimate-body to explode into the lump-masses that became the celestial bodies we see today, and which in turn, are composed of the elements and molecules that were created during the previous universal period(s).

The largest ejected masses became [galactic] force-centres, attracting neighbouring smaller bodies to their vicinity through gravity (magnetism) and creating the galaxies. The kinetic energies in bodies of varying mass caused each to orbit at different velocities and therefore at different orbital radii (Newton) and became galactic satellites.

Galactic satellites become galactic comets when destroyed through impact via orbital precession. These galactic comets are trapped by other galactic satellites and become planetary or lunar satellites.

The relative galactic velocities tell us that they are all moving away from each other, reflecting the ellipsoid (3D) nature of the universe following the original *'Big Bang'*. This movement is the basis of Hubble's law.

Whilst large, the ultimate-body is cold. It comprises all the matter created during the previous universal period and generates sufficient pressure to fuse the matter at its core to create the universe's heaviest elements.

Post *'Big-Bang'*, magnetism is gradually retarding velocities and eventually all universal matter will re-collect back to the ultimate-body; and 'bang', the process starts all over again. This is a never-ending cycle that can continue eternally with no artificial (outside) help, and it complies with all three laws of thermodynamics.

PHILOSOPHIÆ NATURALIS PRINCIPIA MATHEMATICA Revision IV

The '*Big Bang*' occurs when Newton's attraction force $(G.m_p^2/R^2)$ exceeds Coulomb's repulsion force $(k.e^2/R^2)$
Where:
G is Newton's gravitational constant
k is Coulomb's constant
e is the elementary charge
m_p is the mass of a proton
R represents the *diameter* of a proton (two adjacent radii)

Together, these formulas define the mass necessary to balance the attractive (magnetic) and repulsive (electrical) forces:
$m_u = k.e^2 / G.m_p.\varphi + m_p$
Where:
m_u = the ultimate mass (m_u; 'Big Bang' mass)
m_u = >4.7E+48 kg
$N_p = m_u/m_p$ = 2.80364E+75
I.e. there must be more than 2.8E+75 Quanta in the universe

The energy released by the ultimate body of <5.7E+61 J. However, as only $1/63$ of Little-Boy's purely radioactive mass is believed to have exploded, it is assumed here that the explosive energy of the 'Big-Bang' is likely to have been significantly lower, say:

$E_u = e.N_p$ = >4.5E+56 J

If the *mass* of the ultimate-body prior to the explosion is the same as the mass in the universe today (equivalent to 8.784256E+10 Milky Way galactic masses) the average velocity of all galaxies must be equal to $\sqrt{[2.E/m]}$ relative to the centre of the explosion, i.e.:

$v \leq \sqrt{[2.E_u/m_u]}$ = 13,841 m/s

2.7.4 Dark Matter (fiction)

99% of the world's physicists believe that 85% of universal mass is *dark matter*, and that this dark matter comprises sub-atomic particles that cannot be seen, which is the reason it is called *dark*. This has arisen because early in the 20th century, a couple of these physicists (Fritz Zwicky & Jacobus Kapteyn) claimed that Newton's laws of orbital motion predicted that the Milky Way's stars should be thrown into outer space because of centrifugal force based upon its observed star-system population.

The large force-centre at the core of the Milky Way galaxy (Hades) was unknown at that time and was no doubt omitted from these physicists' calculations. However, they should have postulated that there *must* be a mass at the centre of *every* spiral galaxy; because that is how orbits work. I suspect, therefore, that the reason dark matter was invented, is simply that these scientists didn't understand Newton's laws of orbital motion very well.

Given that according to these laws, *only* force-centre mass defines orbital shape, and star-system population *only* defines a force-centre's spin-rate; I find the whole concept of dark matter very difficult to accept because galactic force-centres were unknown 100 years ago and planetary spin theory has only just been resolved.
Moreover, as we now know that electrons do not emit light, we cannot assume that 'black bodies' must necessarily be large enough to trap electrons, they are simply cold enough not to emit electro-magnetic radiation (e.g. light) and could therefore be *any size*.

A galaxy could therefore have a higher star-system population than observed as it may contain cold bodies that cannot be detected. This does not alter the fact, however, that Newton's laws of orbital motion and spin theory *never* predict that galactic systems will fly apart as a result of centrifugal force.

So; why do so many people still believe in dark matter when we should understand these laws much better by now?

The reason that none of this dark matter (sub-atomic particles) has yet been described or discovered is because it doesn't exist.

2.7.5 The Birth of Our Solar System (fiction)

It is currently claimed that our solar system was born from a cloud of gasses that a *'galactic-force'* pushed together into our sun (and planets). The sun then began to spin of its own accord and drag the planets around with it.

It is also claimed that gravity accreted our earth from rocks flying about in the solar system, and that this accretion process generated the heat that currently resides within the planet and that this hot planet slowly became covered with a shallow sea as it cooled down.

There are so many problems with this model it is difficult to know where to begin, but I'll have a go with a few of the obvious ones:

1) Natural hydrogen (H^+) is the only atom (or molecule) that can exist as a gas at <0.1K (in open space). Therefore, our sun was supposedly created from lone protons, which is impossible (see 3) below).

2) A *'galactic-force'* cannot be transmitted through a vacuum.

3) Lone protons (H^+), which constitute 99.3% of all natural hydrogen, cannot accrete irrespective of how hard you push them together. Each proton is applying 2.27E+39 ($1/\varphi$) times more repulsion energy than gravity (magnetism) can apply.

4) Rocks are only created within planets that have sufficient satellite mass to generate the internal frictional heat and consequent volcanic activity. So where did earth's rocky meteorites come from?

5) Without a moon, the earth would have had no internal heat to create volcanoes and earthquakes and hence no atmosphere.

6) More than 99.96% of earth's atmosphere has been generated by its surface life (over the last billion years) and potassium breakdown (over its lifetime). The remainder was created by volcanism which has only been possible since it has acquired a moon, which was long after its *'birth'*.

7) Impact alone does not generate heat, especially in a planet with no atmosphere.

8) Because the early earth existed without a moon, it must have been cold.

9) Spin theory is the only viable answer to the gas planets that can be demonstrated mathematically. So the earth's internal heat can only be generated by its moon and is the reason Venus and Mercury are the only planets in our solar system that have no magnetic field.

10) The first law of thermodynamics tells us that energy cannot be created from nothing. Therefore, the sun cannot spin without an energy causing it to do so. If, as is currently claimed, the sun is dragging its planets around with it, this law also tells us that it should be spinning slower than can be attributed to Hades alone. However, it is actually spinning much faster.

11) If 9) above is correct, the same argument must apply to our sun (and all the stars).

12) Fusion does not generate energy, it requires energy input to work. So, it cannot be possible that our sun evolved from hydrogen (H^+).

13) If stars (including our sun) were created from natural hydrogen (H^+), how can we see their surfaces? Lone protons cannot collect, generate or emit electro-magnetic energy. Electro-magnetic energy can only be generated from proton-electron pairs (H), which are not natural hydrogen atoms.

14) Proton-electron pairs of hydrogen gas (H) are created by fission – they are the last remaining proton-electron pair after converting and removing all the other proton-electron pairs from an atom (as neutrons).

15) The only difference between stars and planets is that stars are brighter due to the orbital systems that generate greater internal heat.

I could continue but it is unnecessary. The current generally accepted model for the universe *must be* incorrect.

Stars are not *created from* hydrogen (H^+) through fusion, they are actually *creating* hydrogen (H) through fission (refer to Chapter 6.5.4).

All universal bodies originally comprised the same matter, that of the ultimate-body. They look and behave according to their orbital systems.

Our solar system is not 4.6bn years old; it is as old as the current universal period (*perhaps* 13.6bn years)

3 Calculation Procedures

A compilation of the mathematical formulas supporting the narrative, including how to use them. This section has been written to simplify their use.

3.1 Energy

The calculation procedures for magnetic and electrical energy have been included in the **Support** section of this book, along with heat, gravity, mass etc. (Refer to Chapters (6.6 to 6.9)

3.1.1 Electrical

Refer to Chapter 6.9

3.1.2 Magnetic

Refer to Chapters 6.7 & 6.8

3.1.3 Electro-Magnetic

Electro-magnetic radiation always travels at the same **velocity**, which we currently refer to as the *speed of light*: c = 299792459 m/s, but is of course, the same speed as *all* electro-magnetic energy.

Knowing this, and that its energy is exactly the same as the kinetic energy in the electron that transferred it, we can calculate its other properties:

Wavelength (λ) is defined by the orbital velocity (v) of the electron transferring it: $\lambda = 2\pi R.c/v = c/f$ {m}
Where: R is the electron orbital (shell) radius

Johannes Rydberg gave us a relationship between wavelength and shell number (n) that actually works:
If the electron temperature (T_1) in shell number 1 (n=1) is known, the electron temperature (T_n) in any shell may be calculated as follows:
$T_n = T_1/n$
For example; the electron temperature at the Bohr [orbital] radius (a_o) which was actually discovered by Rydberg ...
$T = X_R/a_o = 33192.4000063507$ K
Rydberg's constants may be used to calculate the electro-magnetic wavelength and energy generated by an electron in any shell as follows:
$E_n = R_y . T_n/T$
$\lambda_n = \tfrac{1}{2} . (T/T_n)^{1.5} / R_\infty$
Where: X_R (heat transfer coefficient) = 1.75646616508036E-06 K.m

Frequency (*f*) is defined by the orbital period of the electron transferring it: $f = v/2\pi R = c/\lambda$ {Hz}

Amplitude (A) is equal to the orbital radius of the electron transferring it: A = R {m}

Energy (E) may be determined using the modified version of Planck's constant or the orbital velocity (v) of the electron transferring it:
$E = h'/A = \tfrac{1}{2}.m_e.v_e^2$ {J}

A proton develops an operational **Charge** (e') whilst hosting an orbiting electron that maximises at: $e_m = e.\xi_v$
$e' = e . v/v_o$ {C}
This charge is used to generate electro-magnetic radiation it emits and varies with the kinetic energy of the orbiting electron responsible for it.

3.1.4 Potential

Its linear mathematical relationship is: PE = m.g.R = ½.m.v²

In circular orbits, such as atoms, potential energy between particles is *always* twice the kinetic energy in the orbiting satellite, so their mathematical relationships is:

PE = CE = 2.½.m.v² = m.v²

At the speed of light this becomes PE = m.c², which is the true meaning of Henri Poincaré's formula and applies to *orbiting* electrons

3.1.5 Kinetic

Its general mathematical relationship is: $KE = \frac{1}{2}.m.v^2$

The kinetic energy of a satellite orbiting in a circular path is exactly half the satellite's potential energy: $KE = \frac{1}{2}.PE = \frac{1}{2}.m.v^2$

3.2 Orbits

The following Table comprises all the formulas needed to calculate the properties of an orbit.

Sym	Description	units
t	Orbital period	s
R^P	Radius at the orbital perigee	m
θ	Any angle in orbit from apogee	°
R^A	Radius at the orbital apogee	m
m_2	Satellite mass	kg

Table 3.2-1: *Input Data*

Sym	Formula	Description	units
R	$p / [1 - e.\cos(\theta)]$	Orbital radius at θ	m
a	$(R^P + R^A) / 2$	half the major axis of the ellipse	m
e	$-R^P+\sqrt{[\,R^{P2} - 4.a.(R^P-a)\,]} / 2.a$	eccentricity of the ellipse	
b	$\sqrt{[\,a^2.(1-e^2)\,]}$	half the minor axis of the ellipse	m
p	$a.(1-e^2)$	half-parameter (of orbital path)	m
f	R^P	focus distance (orbital perigee)	m
x'	$a - f$	distance from focus to ellipse centre	m
A	$\pi.a.b$	orbital swept area	m²
L	$\pi.\sqrt{[\,2.(a^2+b^2) - (a-b)^2/2.2\,]}$	orbital path length	m
K	t^2/A^3	orbital constant of proportionality	s²/m³

Table 3.2-2: Orbital Shape

Sym	Formula	Description	units
m_1	$\varphi.(2\pi)^2 / G.K$	Force-centre mass	kg
m_2	input	*Satellite mass*	kg

Table 3.2-3: Masses

Sym	Formula	Description	units
v^P	$2.A / t.R^P$	satellite velocity at orbital perigee	m/s
v, v_c	$2.A / t.R$	satellite velocity at θ	m/s
v^A	$2.A / t.R^A$	satellite velocity at orbital apogee	m/s
g^P	$-v^P.v^A / R.(1+e)$	gravitational acceleration at perigee	m/s²
g	$-v.v^A / R.(1+e)$	gravitational acceleration at θ	m/s²
g^A	$-v^P.v^A / R.(1+e)$	gravitational acceleration at apogee	m/s²
F	$-g.m_2$	gravitational force from force-centre	N
F_c	refer to Chapter 3.2.7	centrifugal force in satellite	N
PE	F/R	potential energy between bodies	J
KE	$½.m_2.v^2$	kinetic energy in satellite	J
E	PE + KE	total energy	J
h	$R.v$	constant of motion	m²/s

Table 3.2-4: Orbital Performance

3.2.1 Laws

The following important laws apply to all orbits, irrespective of shape (elliptical or circular):

The key formula for orbital motion originally defined by William Gilbert, and later described as follows by Isaac Newton:
Gravitational Force: $F = G.m_1.m_2 / R^2$
which can be modified to define the following:
Gravitational Energy: $E = G.m_1.m_2 / R$
Gravitational Acceleration: $g = G.m_1/R^2$

Alternative formulas for satellite velocity and gravitational acceleration are:
$v = h/R$
$g = F/m_2$

Throughout any orbit: $E = PE + KE$
where: E is a constant throughout any given orbit
PE is negative and KE is positive

Constant of Proportionality: $K = t^2 / A^2 = (2.\pi)^2 / G.m_1$

3.2.2 Elliptical Orbits

This is nature's perpetual motion machine; it is the first stage of universal energy generation.

A natural (self-energising) orbit is the elliptical path traced by a satellite around its force-centre. Planetary spin (refer to Chapter 3.3) plays no part in this theory.

PHILOSOPHIÆ NATURALIS PRINCIPIA MATHEMATICA Revision IV

3.2.2.1 Input Data

Before calculating the properties of an orbit, we must first identify the input data; i.e. the information required to start the calculation. The input information is usually given as follows because it is the easiest to define:

m_2: satellite mass

t: satellite orbital period

R^P: the perigee radial separation (distance) between the centres of mass of the force-centre and its satellites

R^A: the apogee radial separation (distance) between the centres of mass of the force-centre and its satellites
you need the apogee radial separation (distance) for only one satellite in any orbital system
You can calculate this value for all the other orbits about the same force-centre using a constant of proportionality (K).

G: Newton's gravitational constant; 6.67359232004332E-11 m³/kg/s²

3.2.2.2 Orbital Shape

The properties of an ellipse are well known. Its principal dimensions are shown in Fig 23 and described below

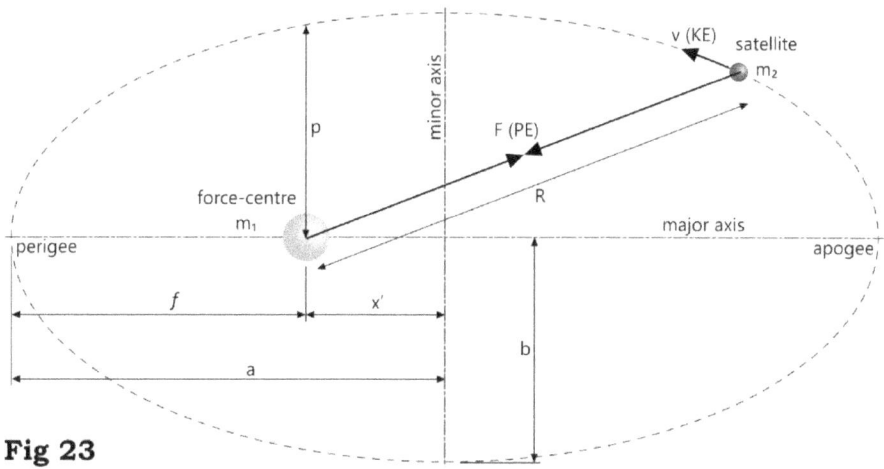

Fig 23

R: radial distance between the satellite and force-centre centres of mass

a: half the major axis of the ellipse

b: half the minor axis of the ellipse

e: eccentricity of the ellipse

p: half-parameter (of orbital path)

f: focus distance from the orbital perigee

x': distance from focus to ellipse centre

v: curvilinear velocity of the satellite

L: circumference of the ellipse (length of orbital path)

t: orbital period; time taken for the satellite to travel around the orbit

θ: angle of 'R' relative to apogee

K: orbital constant of proportionality is the most important orbital feature. It is identical for every orbit encircling the same force-centre

Calculation Procedure

Every orbital system has a unique *constant of proportionality* (K) that is identical for every orbit in the system (a Kepler discovery):
$K = t^2/a^3$
where 't' is the orbital period and 'a' is half the orbit's major axis.

This value for our solar system is:
$K = 2.97491436434708\text{E}{-}19 \ s^2/m^3$
and it applies to all the orbits in our solar system.

For the Milky Way galaxy (refer to Chapter 4.2.1)
$K = 3.35025744599744\text{E}{-}30 \ s^2/m^3$
and it applies to all the orbits in our galactic system.

First we need to use R^P and R^A for just one of the satellites in an orbital system; e.g. the earth (or our sun), from which, we can obtain a value for 'a':
$a = (R^P + R^A) / 2$
$f = R^P$ and $x' = a - f$
$(R^A = x' + a)$

The constant of proportionality (K) applies to a force-centre and is identical for all of its satellites. All remaining 'a' and 'R^A' values can be determined thus:
$a = \sqrt[3]{t^2/K}$
$R^A = 2.a - R^P$

We can now calculate orbital eccentricity:
$e = -R^P + \sqrt{R^{P2} - 4.a.(R^P - a)} \ / \ 2.a$

Half its minor axis can be calculated as follows:
$b = \sqrt{a^2.(1-e^2)}$

Its half-parameter can be found from either:
$p = a.(1-e^2)$ or $p/f = 1+e$

The total swept area of the elliptical orbit is calculated as follows:
$A = \pi.a.b$

The total length of the orbital path can be found using:
$L = \pi.\sqrt{2.(a^2+b^2) - (a-b)^2/2.2}$

The radial distance (R) between a satellite and its force-centre at any point in its orbit:
$R = p / [1 - e.\cos(\theta)]$
refer to Fig 8 for a definition of θ

The velocity can now be found anywhere in its orbit from:
$v = 2.A / t.R$
E.g. the greatest and least orbital velocities are calculated as follows:
$v^P = 2.A / t.R^P$ and $v^A = 2.A / t.R^A$
or from:
$(1+e) = v^{P2} / -g^P.R^P$
only applies to the orbital perigee unless the orbit is circular

The gravitational acceleration between a force-centre and its satellite can be found as follows:
$g = -v^P.v^A / R.(1+e)$ {at apogee and perigee}
$g = -v.v^A / R.(1+e)$ {at any other radius}

The centrifugal velocity of a satellite is not the same as its orbital velocity. It must be modified thus:
$a = \sqrt{[\,^4/_3.\pi\,]}$
$\zeta = \sqrt{[\,(f.\sin(\theta/2)^a + p.\cos(\theta/2)^a) / (f.\cos(\theta/2)^a + p.\sin(\theta/2)^a)\,]}$
$v_c = \zeta.v$
where 'v' is the orbital velocity as calculated above

Therefore, everything you need to know about an orbit can be found from the nearest and farthest distances of a satellite from its force-centre, without Isaac Newton's gravitational constant, 'G' or the mass of either the force-centre (m_1) or the satellite (m_2).

The force-centre's mass (m_1), which is responsible for its orbital shapes, can now be calculated from the above information.

3.2.2.3 Body Mass

You cannot find a satellite's mass simply from the orbital information described above. In fact, you need to *know* the mass of each satellite in the system in order to determine its performance.

But you can find the mass of the force-centre (m_1), using Isaac Newton's gravitational constant (G) together with the constant of proportionality (K). This means that all the information about every orbit in a given system can be established by using Newton's 'G' only once in one of the orbits.

Calculation Procedure

Force-centre mass may be calculated from the first orbital analysis thus:
$m_1 = (2.\pi)^2 / G.K$
or you may use Isaac Newton's famous formula if you prefer:
$m_1 = -g.R^2 / G$
This is the only time you need to apply Newton's gravitational constant (G); i.e. m_1 (and 'K') remains the same for all other satellite orbits.

You would normally enter a known value for the satellite mass (m_2), but if it is unknown, you can use the following formulas where the gravitational acceleration (g_s) of the satellite at a specified radius (r_s) is known:
either by using Isaac Newton's gravitational constant
$m_2 = -g_s.r_s^2 / G$
or from the following formula if you've forgotten 'G':
$m_2 = m_1 . g_s/g^P . (r_s/R^P)^2$

3.2.2.4 Satellite Performance

Satellite performance constitutes the forces and energies that vary around an orbit. These can be seen in Fig 24 and described below.

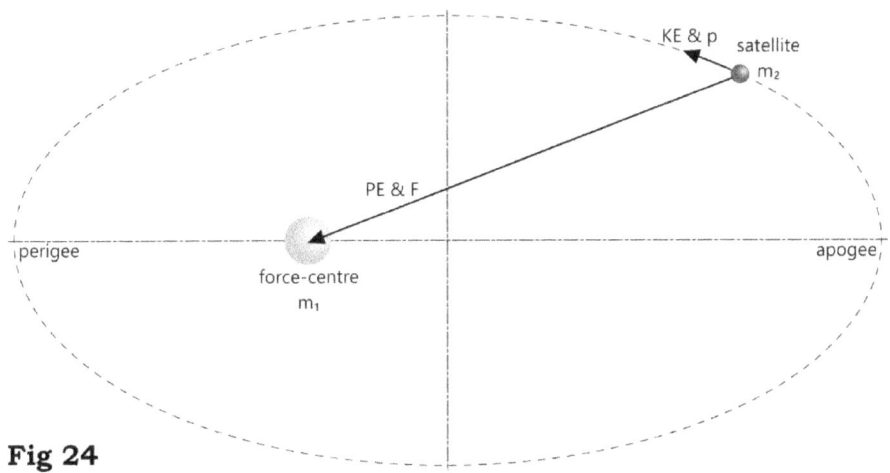

Fig 24

F: gravitational force between the force-centre and the satellite (-ve)

PE: potential energy between the force-centre and the satellite (-ve)

KE: kinetic energy in the satellite (+ve)

E: total energy in the orbital system

p: satellite momentum, which varies with orbital velocity

h: Newton's constant of motion (satellite momentum without the mass component)

Gravitational force may be calculated using Isaac Newton's formula:
$F = -G.m_1.m_2 / R^2$
or an alternative formula:
$F = -g.m_2$

Centrifugal force in the satellite may be calculated as follows:
F_c (refer to Chapter 3.2.7)

Potential energy can be calculated by multiplying the radial distance between the centres of mass of the two bodies by the gravitational force between them:
$PE = F.R$ (where F is always negative)

Kinetic energy in the satellite varies with orbital velocity and can be calculated thus:
$KE = \frac{1}{2}.m_2.v^2$

Total energy is the sum of the two energies above:
$E = PE + KE$

Satellite momentum is calculated thus:
$p = m_2.v$

Newton's constant of motion may be calculated thus:
$h = R.v$

3.2.3 Circular Orbits

The magnetic potential energy (PE$_m$) between a force-centre and its satellite(s) applies to all satellites, irrespective of an electrical potential energy (PE) between the same force-centre and its satellite(s).

PE$_m$ is 'φ' times PE

The potential energy between a force-centre and a satellite orbiting in a circular path is *always exactly* twice the satellite's kinetic energy. This means that the potential energy in an electron is; PE = 2.½.m.v^2 = m.v^2

The principal features of an orbit are described below (and Fig 23):

Minor & Major Axes: are identical

Perigee & Apogee: are identical

R: the distance between the centres of the satellite and its force-centre, which remains constant throughout a satellite's orbit

F: the magnetic or electrical force imposed on the satellite by its force-centre, which remains constant throughout the satellite's orbit

v: the curvilinear velocity of the satellite which remains constant throughout a satellite's orbit

3.2.3.1 Input Data (common)

Before calculating the properties of an orbit, we must first identify the input data; i.e. the information required to start the calculation.

The input information common to both types of circular orbit (magnetic and electrical), is usually given as follows because it is the easiest to define:

e: orbital eccentricity = 0
(for all orbital shells)

G: Newton's gravitational constant; 6.67359232004332E-11 $m^3/kg/s^2$
(refer to Chapter 6.11.2)

3.2.3.2 Orbital Shape

The properties of an ellipse are well known. Its principal dimensions are shown in Figs 23 & 25 and described below

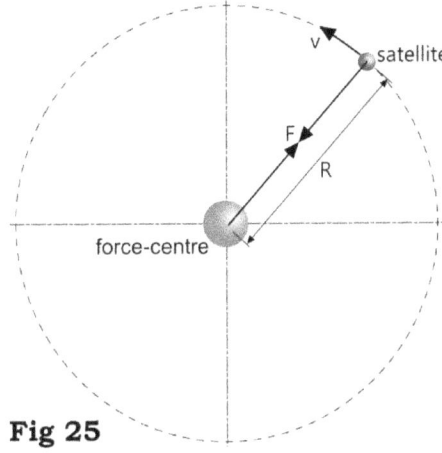

Fig 25

R: radial distance between the satellite and force-centre centres of mass

a = b = p = f = R: orbital radius

e: eccentricity of the ellipse = 0

x': distance from focus to ellipse centre = 0

v: curvilinear velocity of the satellite

L: circumferential length of the circular orbital path

t: orbital period; time taken for the satellite to travel around the orbit

K: orbital constant of proportionality is the most important orbital feature. It is identical for every satellite orbiting the same force-centre

3.2.3.3 Magnetic

The *magnetic* orbit refers to gravitational orbits such as man-made communication satellites.

Input Data (specific)

Before calculating the properties of an orbit, we must first identify the input data; i.e. the information required to start the calculation. The input information is usually given as follows because it is the easiest to define:

m_1: force-centre mass

m_2: satellite mass

v: satellite velocity

Calculation Procedure (Orbital Shape)

First, we need to find the orbital radius, which is defined by the velocity of the satellite:
$R = G.m_1/v^2$

The magnetic attraction holding onto the satellite can be calculated as follows:
$g = v^2/R$

The orbital period is the time taken for a satellite to complete a single orbit:
$t = 2.\pi.R/v$

We now know that in a circular orbit:
$e = 0$ and $a = b = p = f = R$
and the orbital area and the orbital path length are both the same as those for a circle in circular orbits:
$A = \pi.R^2$ and $L = 2.\pi.R$

The constant of proportionality (K) remains constant for all satellites orbiting the same force-centre. It may be calculated thus:
$K = t^2/a^3$
where 't' is the orbital period and 'a' is half the orbit's major axis (R).

m_1 can now be confirmed using 'K' and 'G'.

Body Mass

The satellite and force-centre masses are included in the input data for circular orbit calculations. However, it is useful to know that m_1 can be confirmed as follows:

Calculation Procedure

Force-centre mass may be calculated from the first orbital analysis thus:
$m_1 = (2.\pi)^2 / G.K$

or you may use Isaac Newton's famous formula if you prefer:
$m_1 = g.R^2 / G$

We already know 'm_2' from the input data.

Satellite Performance

Satellite performance constitutes the constant orbital forces and energies. These can be seen in Fig 26 and described below.

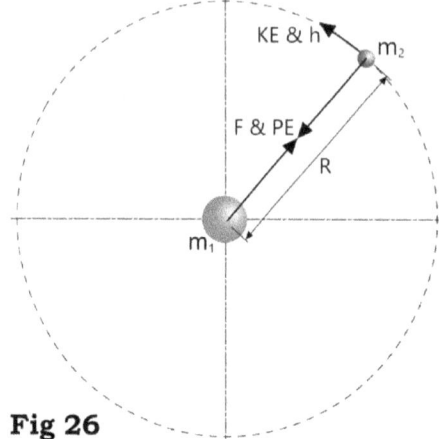

Fig 26

F: attractive force between the force-centre and the Satellite (-ve)

PE: attractive energy between the force-centre and the Satellite (-ve)

KE: kinetic energy in the satellite (+ve)

E: total energy in the orbital system

p: satellite momentum, which varies with orbital velocity

h: Newton's constant of motion (satellite momentum without the mass component)

Calculation Procedure:

Attractive force may be calculated as follows:
$$F = -G.m_1.m_2 / R^2$$
or an alternative formula:
$$F = -g.m_1$$

Centrifugal force in the satellite may be calculated as follows:
$$F_c = m_2.v^2/R$$

Potential energy can be calculated by multiplying the radial distance between the centres of mass of the two bodies by the gravitational force between them:
$$PE = F.R = -2.KE$$

Kinetic energy in the satellite varies with orbital velocity and can be calculated thus:
$$KE = \tfrac{1}{2}.m_2.v^2 = -\tfrac{1}{2}.PE$$

Total energy is the sum of the two energies above:
$$E = PE + KE$$

Satellite momentum is calculated thus:
$$p = m_2.v$$

Newton's constant of motion may be calculated thus:
$$h = R.v$$

3.2.3.4 Electrical

The *electrical* orbit refers to atomic orbits only.

Input Data (specific)

Before calculating the properties of an orbit, we must first identify the input data relating to atomic particles; i.e. the information required to start the calculation. The input information is usually given as follows because it is the easiest to define:

m_p: proton mass = 1.67262163783E-27 kg

m_e: electron mass = 9.1093897E-31 kg

T: temperature; magnitude of electromagnetic energy (heat) absorbed by the electron

X: Heat constant; 6.9353271647894E-09 K.s²/m²
(refer to Chapter 6.6)

X_R: Heat constant; 1.75646616508036E-06 K.m
(refer to Chapter 6.6)

φ: coupling ratio = 4.40742111792335E-40

Calculation Procedure (Orbital Shape)

First, we need the relative electrical charge:
$RAC = e/m_e$

The velocity of an orbiting electron can now be calculated as follows:
$v = \sqrt{[T/X]}$

The orbital radius of an orbiting electron is dependent upon its velocity, and may be calculated thus:
$R = G.m_p / \varphi.v^2$ or $R = X_R/T$

We now know that in a circular orbit:
$e = 0$ and $a = b = p = f = R$

The potential acceleration in the proton-electron pair is:
$g = v^2/R$

The orbital period is the time taken for an electron to complete a single orbit:
$t = 2.\pi.R/v$

We need the separation between any two electrons in an orbit, which will be the same as 2.R. This is calculated for any orbital shell as follows:
$d = \pi . (4.\pi.R^2) / 2.(2.\pi.R)$ {the arc length}
$\ell = 2.R.\mathrm{Sin}(d / 2.R)$ {half the straight-line distance}

The orbital area and the orbital path length are both the same as those for a circle in circular orbits:
$A = \pi.R^2$
$L = 2.\pi.R$

An important fact to remember about the constant of proportionality (K) is that it remains constant for all electrons irrespective of shell number (radius) or velocity (temperature).
It may be determined as follows:
$K = t^2/a^3 = 0.15587874533403$ {s^2/m^3}
where 't' is the orbital period and 'a' is half the orbit's major axis.

The above constitutes everything you need to determine the size of every electron shell in an atom. Each shell can hold up to two identical electrons, which is confirmed by 'ℓ' above being the same as 'R'.
m_p can now be confirmed using 'K' and 'G'.

Body Mass

The proton and electron masses are included in the input data for atomic shell calculations. However, it is handy to know that m_p can be confirmed as follows.

Calculation Procedure:

Force-centre mass may be calculated from the first orbital analysis thus:
$m_p = \varphi.(2.\pi)^2 / G.K$

or you may use Isaac Newton's famous formula if you prefer:
$m_p = \varphi.g.R^2 / G$

This is the only time you need to apply Newton's gravitational constant (G); i.e. m_p (and 'K') remains the same for all other electron orbits, irrespective of electron velocity and shell number.

The coupling ratio (φ) is needed in the above formulas because it is the electrical charge that is holding the electron to the proton, not magnetism.

We already know 'm_e' from the input data.

Electron Performance

Electron performance constitutes the constant orbital forces and energies. These can be seen in Fig 27 and described below.

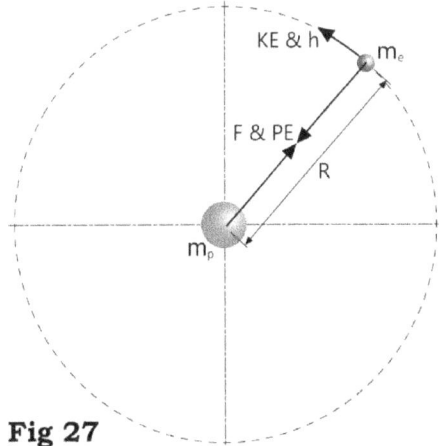

Fig 27

F: attractive force between the proton and the electron (-ve)

PE: attractive energy between the proton and the electron (-ve)

KE: kinetic energy in the electron (+ve)

E: total energy in the orbital system

p: electron momentum, which varies with orbital velocity

h: Newton's constant of motion (electron momentum without the mass component)

Together with the previously calculated information (refer to Chapters 3.2.3.1 to 3.2.3.3), the above provides everything you need to resolve electron performance in atoms.

Calculation Procedure

Electrical attractive force may be calculated as follows:
$F = -g.m_e$

Centrifugal force in the orbiting electron may be calculated as follows:
$F_c = m_e.v^2/R$

Potential energy can be calculated by multiplying the radial distance between the centres of mass of the two bodies by the gravitational force between them:
$PE = F.R = 2.KE$

Kinetic energy in the electron varies with orbital velocity and can be calculated thus:
$KE = \frac{1}{2}.m_e.v^2 = \frac{1}{2}.PE$

Total energy is the sum of the two energies above:
$E = PE + KE$

Electron momentum is calculated thus:
$p = m_e \cdot v$

Newton's constant of motion may be calculated thus:
$h = R \cdot v$

You may check the validity of your results as follows:

Calculate the magnetic properties of the atomic orbit using the calculation procedure described in Chapter 3.2.3.3 and compare the calculation results with those in this Chapter (3.2.3.4)

All the following ratios must equal the coupling ratio (φ) ...
g_m/g; F_m/F; Fc_m/Fc; PE_m/PE; KE_m/KE; E_m/E

... and these must equal the square root of the coupling ratio ($\sqrt{\varphi}$):
v_m/v; h_m/h

... if your calculation results are correct.

3.2.4 Force-Centre Mass

As is demonstrated below, it is easy to establish an accurate force-centre mass (m_1) if the major axis ($R_P + R_A$) of just one satellite orbit is known accurately.

$a = \frac{1}{2}.(R_P + R_A)$

Where:

R_P is the radial distance between the centres of the satellite and its force-centre at the satellite's perigee

R_A is the radial distance between the centres of the satellite and its force-centre at the satellite's apogee

$K = t^2/a^3$

Where:

t is the satellite's orbital period

$m_1 = (2.\pi)^2 / G.K$

Where:

G = Newton's gravitational constant

Newton's gravitational constant (G) is not needed for the calculation of any other satellite in the same system (refer to Chapter 3.2.2.2).

3.2.5 Planetary Mass

Using Galileo's measurement of gravitational acceleration, Kepler's law of equal orbital time and swept-Area and Newton's laws of orbital motion we can calculate the mass of a distant satellite from its orbital deviation whilst passing another body of known mass.

By dropping our ball (or rolling it down a slope) and measuring the time it takes to cover a known height, we can find the gravitational acceleration on the surface of our planet, an accurate value for which is now available:

The earth's mean radius: r = 1.496E+11 m
Newton's gravitational constant: G = 6.67359232E-11 m^3/kg/s^2
The earth's mean gravitational acceleration at sea level:
g = 9.80663139 m/s^2

Using Newton's formula; $g = G.m_2/R^2$, we can now find the earth's mass (m_2).

As we know the earth's orbital shape and its velocity (v) at any point within it, we can accurately determine its centrifugal force (F_c) at its perigee for example, from:
$F_c = f/p . m_2.v^2/R$

and because we know that the gravitational force between the earth and sun is exactly the same as the earth's centrifugal force (i.e. they balance each other exactly), we can establish the mass of our sun (m_1) from:
$F_c = G.m_1.m_2 / R^2$
Alternatively, our sun's mass could be determined by using the constant of proportionality from any of its satellite orbits ($m_1 = (2.\pi)^2 / G.K$)

Knowing the mass of our sun and the earth along with the orbital shape of the other bodies in our solar system, we can calculate their masses by comparing force-triangles, one from a planet's theoretical orbit and the other as it passes close to one of which the mass is known, using its orbital deviation and the formulas:
$g = G.m_1/R^2$ & $m_2 = F.R^2 / G.m_1$

So, from measuring the time it takes to drop a ball anywhere on our planet, along with the observed orbits for all the planets and moons in the solar system, Newton has given us all we need to calculate the mass of every celestial body in the solar system.

3.2.6 External Influences

An external influence on a satellite is defined by the gravitational pull between itself and another body.

$PE = G.m_1.m_2 / R$

where:
PE = the potential energy between m_1 and m_2
G = Newton's gravitational constant
m_1 = the mass of the body pulling the satellite out of its orbit
m_2 = the mass of the satellite
R = the radial separation between the centres of the two bodies

3.2.7 Centrifugal Force

Any orbiting mass will be subject to a centrifugal acceleration (a). Christiaan Huygens gave us the relationship between this and its velocity
$$a = v^2 / R$$
where 'v' is its curvilinear velocity and 'R' is its radius of motion.

However, the above velocity (v) must be modified for elliptical orbits (v_c) as shown in Fig 8, and looks like this:

$$a = \sqrt{[^4/_3 . \pi]}$$
$$v_c = \zeta . v$$

Where: $\zeta = \sqrt{[\,(f.\sin(\theta/2)^a + p.\cos(\theta/2)^a) / (f.\cos(\theta/2)^a + p.\sin(\theta/2)^a)\,]}$

Centrifugal force, which is equal and opposite to centripetal force, is calculated thus: $F_c = m_2 . v_c^2 / R$
which may be simplified at the orbital extremes as follows:
@ the perigee (perihelion) of an ellipse; $F_c = F . f/p = F / (1+e)$
@ the apogee (aphelion) of an ellipse; $F_c = F . p/f = F . (1+e)$

3.2.8 Station-Keeping

The variation in a satellite's PE & CE as it is pulled off course, may be defined by comparing its gravitational (g) and centrifugal (a) acceleration at any given angle through its orbit (θ), by altering its orbital radius.

g = gravitational acceleration (refer to Chapter 3.2.2.2)

a = $(v.\zeta)^2 / R$
where:
v = satellite curvilinear velocity (refer to Chapter 3.2.2.2)
ζ = factor (refer to Chapter 3.2.7)

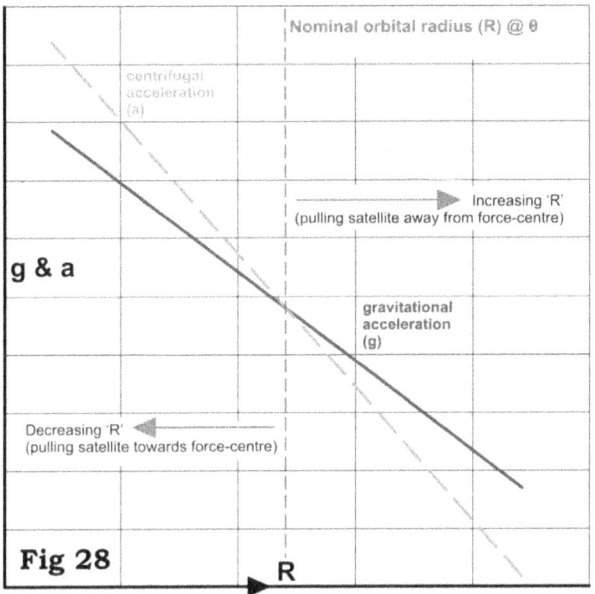

Fig 28

Plotting 'g' and 'a' will result in the following (Fig 28):

As R increases g>a so the satellite will return to its orbit

As R decreases a>g so the satellite will return to its orbit

PHILOSOPHIÆ NATURALIS PRINCIPIA MATHEMATICA Revision IV

3.3 Spin

This is nature's energy generator; it is the second stage of universal energy production.

Planetary spin is a fundamental part of the laws of orbital motion but today, it is believed impossible to calculate. Luckily, this is not the case.

Whilst it was not addressed by Isaac Newton - because he did not have the information required to resolve it – it can be solved using his laws of orbital motion.

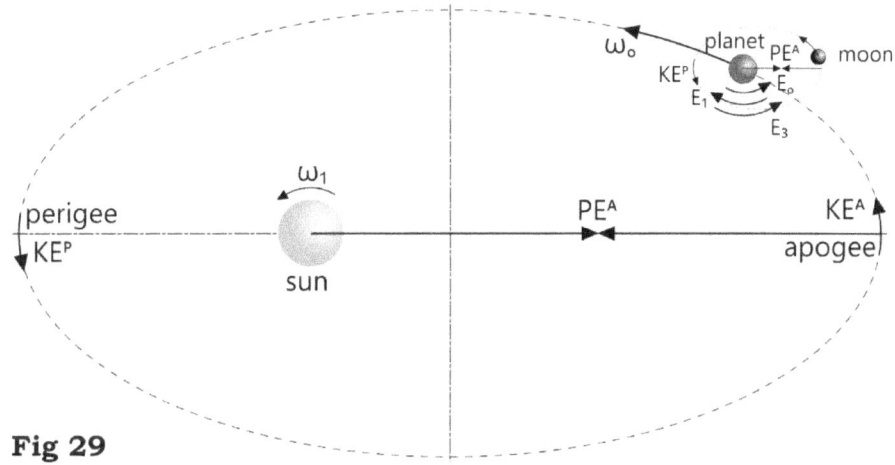

Fig 29

Fig 29 shows the energies that generate spin in each orbiting and/or orbited body. Only orbiting and orbited bodies induce spin in each other. I.e. the sun induces spin in the earth, as does Jupiter and its moons, but Jupiter does not induce spin in Mercury or Neptune.

Spin is induced in orbiting and orbited bodies by the rotational and orbital energies (KE and JE {spin energy}) via the energy that binds them together (PE).

Spin direction: I shall define retrograde as positive and prograde as negative, which defines the orientation of E_0 to E_3 below. This definition may be reversed according to preference.

Spin is induced in one body by the kinetic (orbital and/or rotational) energies in another, via the potential (gravitational) energy between them.

The energies that generate spin in a **satellite** are listed below:

The energy driving a **satellite**'s spin is calculated as follows:
$E_2 = E_1 - E_3 - E_0$

E_0 is the energy developed from the **satellite**'s natural rotation as it orbits around its **force-centre** (star)
$E_0 = \frac{1}{2}.J.\omega_0^2$

E_1 is the rotational energy induced in the **satellite** by the rotational kinetic energy in its **force-centre** (star)
$E_1 = \delta KE \cdot (r/R)^2$
Where; $\delta KE = \Sigma KE^P - \Sigma KE^A$, 'r' is the radius of the **satellite** and 'R' is the radial distance between the centres of the **force-centre** (star) and the **satellite**.

E_3 is the angular velocity induced in a **satellite**'s mass by its orbiting **sub-satellite**(s)
$E_3 = \Sigma(KE^P + PE^A) \cdot \text{Sign}[\text{Cos}(\theta)]$
Where:
θ is the relative tilt between **satellite** and **sub-satellite** (lunar) orbital planes.
$\Sigma(KE^P + PE^A)$ is the sum of such energies of each **sub-satellite** (moon) orbiting the planet.

The angular velocity of the **satellite** is calculated thus:
$\omega_2 = \sqrt{[2.E_2 / J_2]}$

This calculation method is as accurate as Newton's own laws of motion and is essentially an extension of them. Therefore, not only is planetary spin predictable, it is both simple and accurate, demonstrated by the fact that there now exists a simple calculator for this calculation (http://calqlata.com/proddetail.asp?prod=00085).

Input Data

Before embarking on the calculation procedure, we must first identify the input data, all of which can be found from observation and/or Newton's laws of orbital motion:

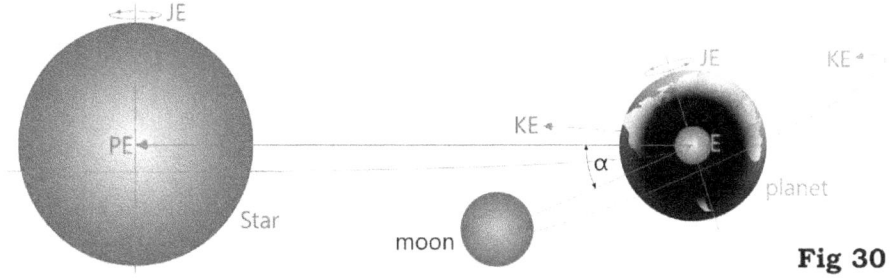

Fig 30

α is the angle between the two orbital planes (Fig 30)

t is orbital period (in seconds). The time taken for the **satellite** to complete one full orbit

r is the outside radius of the **satellite**

Δ is a radial modifier that represents the density variation inside a **satellite**

> Δ < 1 means the core density is greater than its surface density
> Δ = 1 means that the entire planet is homogeneous
> Δ > 1 indicates that the planet is being pulled into a local orbit (e.g. Pluto)
> The greater the Δ value (up to 1.0) the lower the density variation.

R^P distance between a **satellite** and its **force-centre** at its orbital perigee

R^A distance between a **satellite** and its **force-centre** at its orbital apogee

KE^P kinetic energy in a **satellite** at its orbital perigee

KE^A kinetic energy in a **satellite** at its orbital apogee

ΣKE^P sum of the kinetic energies in all the **sub-satellites** at their orbital perigee

ΣKE^A sum of the potential energies in all the **sub-satellites** at their orbital apogee

Note: It is important to apply appropriate -ve and +ve polarity to the above 'KE' values according to their prograde or retrograde directions.

Calculation Procedure

This calculation procedure identifies the rotary spin in a **satellite** (E_2) due to its own orbit (E_0), its **force-centre** (E_1) and its own **sub-satellites** (E_3)

Spin polarity; defines direction with respect to **sub-satellite** axis tilt:
$\psi = \text{Sign}(\text{Cos}(\alpha))$ {1 or -1}

Average orbital radius of the **satellite**:
$R_{ave} = \frac{1}{2}.(R^P + R^A)$

Polar moment of inertia of the **satellite**:
$J = \frac{2}{5}.m_2.(\Delta.r)^2$

The difference between the kinetic energy in the **satellite** at its orbital perigee and its orbital apogee:
$\delta KE = \Sigma KE^P - \Sigma KE^A$

The sum of all the kinetic energies in the **sub-satellites** at their orbital perigee and their potential energies at their orbital apogee:
$E^s = \Sigma KE^P + \Sigma PE^A$

The rotational velocity in the **satellite** about its axis of spin:
$\omega_0 = 2\pi/t$

The rotational energy in the **satellite** due to 'ω_0':
$E_0 = \frac{1}{2}.J.|\omega_0|\omega_0$
'$|\omega_0|\omega_0$' is the same thing as ω_0^2 multiplied by the sign (polarity; -1 or +1) of ω_0 and therefore gives E_0 direction

The rotational energy in the **satellite** due to the spin in its **force-centre**:
$E_1 = \delta KE.(r/R_{ave})^2$

The total rotational energy in the **satellite**:
$E_2 = E_1 - E_0 - E_3$

The rotational energy in the **satellite** due to its orbiting **sub-satellites**:
$E_3 = \psi.E^s$ (refer to Chapter 4.3.4)

Satellite spin velocity (radians per second) may be calculated as follows:
$\omega_2 = \text{Sign}(E_2) . \sqrt{[2.|E_2| / J]}$
'$|E_2|$' is the positive value of E_2 and 'Sign(E_2)' is its polarity giving ω_2 direction

3.3.1 Polar Moment of Inertia

Polar moment of inertia (J) is calculated thus:
$J = \frac{2}{5}.m.r^2$
where:
'm' is the body's mass
'r' is its outside radius

The radial modifier (Δ) is applied thus:
$J = \frac{2}{5}.m.(\Delta.r)^2$ (Δ < 1)

The greater the Δ value the lower the density variation;

Δ < 1 means the core density is greater than its surface density (normal situation)

Δ = 1 means that the entire planet is homogeneous

Δ > 1 indicates that the body is being pulled into a local orbit

3.3.2 Earth's Core

A planet's core will be forced to rotate at a different rate to its outer mantle if the conditions are correct. This is due to the kinetic energy in a large moon inducing rotational energy (E_0) throughout the planet in one direction, and the spin energy in its force-centre (E_1 - E_3) inducing spin in the planet's core in the opposite direction.

When applied to the earth's core, spin theory can show us the difference between the angular velocity of the earth's inner core and its mantle:

Earth's mass; m_E = 5.9662986112E+24 kg
Earth's polar moment of inertia; J_E =1.08209548E+37 = kg.m² #
Radius of earth's core; r = 1215000 m
Density of earth's core; ρ = 7870 kg/m³
Volume of earth's core; V = 7.51307013637E+18 m³
Mass of earth's core; m = 5.91278619733E+22 kg
Mass of mantle and outer core; m_m = 5.90539190574E+24 kg
Polar Moment of Inertia:
Earth's core; $J = \tfrac{2}{5}.m.r^2$ = 3.49144112166E+34 kg.m²
Earth's mantle; $J_m = J_E - J$ = 1.07860404E+37 kg.m²
Rotational Energies:
E_0 = -2.14473244632E+23 J #
$E_1 - E_3$ = 2.8770397039E+28 J #
Rotational Velocities:
Earth's core; $\omega = \sqrt{[2.E_0/J]}$ = -3.50509019131E-06 ^{c}/s
Earth's mantle; $\omega_m = \text{sign}(E_1+E_3).\sqrt{[2.(E_1+E_3)/J_m]}$ = 7.3039350764E-05 ^{c}/s
$\delta\omega = \omega + \omega_m$ = **6.95342605725E-05 ^{c}/s**

Table 4.3.5-1

3.3.3 Earth's Magnetic Field

There are two competing energies driving the spin in the earth's core and its mantle:
-E_0 (-2.1447324E+23 J) is the sun's energy driving the core
E_1-E_3 (2.87704E+28 J) is the moon's energy driving the mantle (and core)

The polar moments of inertia:
Core: J = 3.49144112166E+34 kg.m²
Mantle: J_m = 1.07860404E+37 kg.m²

The angular velocities:
Core: ω = Sign(E_0) . $\sqrt{[2.|E_0|/J]}$ = -3.50509019131E-06 ᶜ/s
Mantle: ω_m = Sign(E_1-E_3) . $\sqrt{[2.|E_1-E_3|/J_m]}$ = 7.3039350764E-05 ᶜ/s

The differential angular velocity:
$\delta\omega$ = ω + ω_m = **6.95342605725E-05 ᶜ/s**

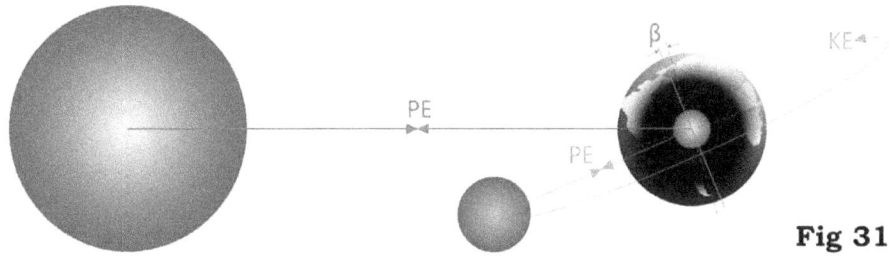

Fig 31

The angular tilt (β) between the two axes can be calculated thus:
β = sign(ω/ω_m) . ½.$\sqrt{[|A\sin(\omega/\omega_m)|]}$
 = 0.109553685228394 radians (**6.27696379369167°**)

3.3.4 Magnetic Reversal

N/A

3.3.5 No Moon!

The rotational energy formula (Chapter 5: $E_2 = E_1 - E_3 - E_0$) then becomes:
$E_2 = E_1 - E_0$
In which;
$E_1 = 3.208E+23$ J
$E_0 = 2.14479E+23$ J
$E_2 = E_1 - E_0 = 3.208E+23 - 2.14473245E+23 = 1.06321E+23$ J
$E_3 = 0$ (zero)
as opposed to 2.87709E+28 J (with its moon)

'*No-moon*' will reduce the earth's internal heat and its magnetic field by a factor of $\approx 3E+05$ ($\delta\omega \approx 0$).

3.3.6 Hades

What do we know about Hades?

It comprises the same matter as the ultimate-body. It has no force-centre of its own, so it generates no internal friction. It is cold, i.e. it emits insufficient electro-magnetic energy to detect it. It comprises the same matter as all other universal bodies, giving it a density (ρ) of **5300 kg/m³**

If the Milky Way's orbital eccentricity is 0.016 and *if* it takes 230 million years for our solar system to orbit, either or neither of which may be correct, Hades' mass is **1.76572019E+41 kg**

If the above mass and density are correct, Hades' diameter is **4.0E+12 m** from: $r = \sqrt[3]{[\ 3.m_1\ /\ 4.\pi.\rho\ ]} \approx 1.74966E+12$ m

Hypothetically: *If* we were to define a *photon* as an electron travelling at the speed of light and a black-hole as a black body large enough to trap *photons* (refer to Chapter 6.2.2), then; $PE_{bh} \geq KE_e$

Where; PE_{bh} is the gravitational energy at the surface of a black hole and KE_e is the kinetic energy in a photon:
$PE_{bh} = G.m_1.m_2/R = 6.1350483563E-12$ J
$KE_e = \frac{1}{2}.m_2.c^2 = 4.093555584E-14$ J
Therefore, Hades has more than enough gravitational energy to trap a *photon*.

The minimum size for black-hole comprising mostly iron, may be calculated as follows:
$E = G.\ m_1.\cancel{m_2}/R = \frac{1}{2}.\cancel{m_2}.c^2$
Given that; $R = \sqrt[3]{[3.m_1\ /\ 4.\pi.\rho]}$
$m_1 = \sqrt{[3.c^6\ /\ 32\pi.\rho.G^3]} =$ **9.623785516E+37 kg**
Therefore, Hades is larger than the minimum sized black-hole.

However, as photons do not exist, the above *black-hole* is hypothetical.

If the Milky Way contains 100bn star-systems similar to our solar system, Hades rotates at **2.012413E-07 c/s**

If the Milky Way contains 10bn star-systems similar to our solar system, Hades rotates at **6.36381E-08 c/s**

I.e. despite the fact that we cannot see, hear or feel Hades, Newton's laws can tell us an awful lot about it.

Despite the fact that Hades spins on its axis it generates no heat because it has no force-centre of its own.

When calculating the spin in a force-centre that does not follow an orbital path, such as Hades:
Set E_0 & E_1 to zero

When calculating the spin in a sub-satellite that has no sub-sub-satellites, such as our moon:
Set E_3 to zero

Having demonstrated that our sun obeys Newton's laws in orbit around the Milky Way and having established Hades' mass we can now determine its spin rate.

First, however, we need to have a stab at establishing its density. We already know that the largest stable, naturally occurring element in the universe is iron, which is the heaviest atom generated through fusion in the core of a galactic force-centre. We also know that negative magnetic charge (gravity) is 4.407E-40 times that of the positive electrical charge. Using core pressure theory, it has been established that the pressure in the core of Hades is 4E-05 times that required to compress an iron atom and thereby increase its density (refer to Chapter 2.7.2). It is therefore a safe bet that Hades has a density that cannot exceed iron, but is most likely the same as all other universal matter; ≈5300 kg/m³.

Using spin theory, it has been possible to establish the following about Hades:

mass of Hades (m)	1.7657E+41	Kg
density of Hades	≈ 5300	Kg/m³
radius of Hades (r)	≈ 1.996116117E+12	m
moment of angular inertia of Hades ($J_H = \frac{2}{5}.m.r^2$)	2.814190398E+65	kg.m²
energy of the sun used to rotate Hades ($E_S = KE_P - PE_A$)	4.3781615644E+40	J
angular kinetic energy in Hades ($E_H = \frac{1}{2}.J_H.\omega^2$)	4.3781615655E+51	J
number of Suns rotating Hades (N)	1.0E+11 #	
angular velocity of hades ($\omega = \sqrt{[2.E_S.N / J_H]}$)	1.76394139223E-07	ᶜ/s
Table 3.3.6-1: Hades Angular Velocity		

NASA estimate

You will note, that contrary to popular belief, a *black hole* is simply a slow-spinning ball of matter too cold to emit significant levels of electro-magnetic energy.

When calculating the rotational velocity of Hades, it is important to remember that we know neither the number of star-systems nor do we know the orbital radii of any but our own. Therefore, we can only estimate the total satellite induced spin energy based upon our own solar system multiplied by the number of estimated stars-systems. It is currently estimated that the Milky Way galaxy contains 100bn star-systems. As we have no way of estimating this value with any degree of accuracy, we may try an alternative value to see what effect it would have on Hades' spin-rate. As I believe that 100bn is excessive, I have provided an alternative calculation for 10bn star-systems similar to our own (refer to Chapter 4.3.6)

3.4 Core-Pressure

Core pressure means the internal pressure within any mass (even a lump of metal) due to the same gravitational forces as those attracting satellites to their force-centres.

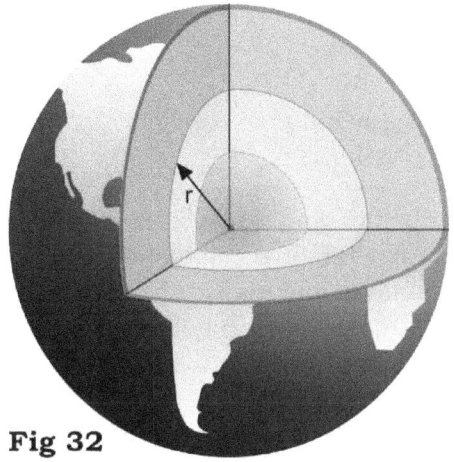

Fig 32

This pressure can be used to define the internal structure of a planet or star, neither of which will be homogeneous. For example, the earth's iron core will have a higher density than its outer core or mantle material. Your structural calculation must generate the same total mass and polar moment of inertia already determined through spin theory (refer to Chapter 3.3).

A simple calculation procedure exists that will provide a fairly reliable result.

The only variable needed when calculating core pressures is a body's 'Δ' value.

The simplest practical calculation procedure is to break the planet up into spherical layers and estimate (iterate) the density and radius of each layer (Fig 32). When the calculation reproduces the correct total mass and polar moment of inertia you will have a representative internal construction.

Newton's gravitational formula $F = G.m_1.m_2 / R^2$ provides all we need to calculate the pressure anywhere inside a body. When used together with the body's Δ value (refer to Chapter 4.3.1), we can determine the internal pressure anywhere inside a body with a complex (layered) structure, such as a planet or star.

But first of all, we need to modify the formula slightly:
$m_2.g = F = G.m_1.m_2 / r^2$
$g = G.m_1 / r^2$
i.e. at any radius within a spherical body, the gravitational acceleration is g based upon the mass of material inside radius r.
Knowing 'g' at any radius 'r', if the total body mass = m
m_1 is the mass inside 'r', and
m_2 is the mass outside 'r' ($m_2 = m - m_1$)
Given that pressure: $p = F/A = m.g / A$
And using the corrected value for Newton's gravitational constant
(G = 8.38628344228057E-10 m³/s²/kg): $p = G.m_2.m_1 / A^2$

For non-homogeneous bodies, they must be broken into layers, each of which will have a representative density. After the final calculation the sum of the polar moment of inertia of each layer must equal the polar moment of inertia of the body ($J = ⅖.m.(\Delta.R)^2$) and the sum of the mass of each layer must equal the body's total mass (m). Whilst this is an iterative process it is quite simple and gives an accurate result.

With regard to the calculation for the earth (refer to Chapter 4.4), reducing the core density to that of iron as we understand it, changes the upper mantle density by only a small percentage (<1%). Given the relatively low mass of the earth's core when compared with that of its mantle along with the atomic compressibility of 'φ' (4.407E-40) the core density is almost certainly close to that of iron (7870 kg/m³) and the suggested upper mantle density (close to that of water) must be correct.

This calculation procedure (see below) has revealed that the earth's upper mantle density is significantly less than the crust material it is supporting. It also tells us that the internal pressure is insufficient to support the overlaying crust. However, by applying 'PV=RT', we can establish the amount of heat at the upper-mantle crust interface.

What do we know about the inside of the earth?
1) Its core is made of iron #
2) Its surface comprises mostly water
3) Its inner core radius is ≈1215000m
4) Its outer core radius is ≈3470000m
5) Its upper mantle radius is ≈3463000m *(6km crust and 2km water depth)*
6) Its outer radius is 6371000m
7) Its total mass is 5.96451976771E+24 kg *(Chapter 4.2.2)*
8) Its polar moment of inertia is 1.08209548E+37 kg.m² *(Chapter 4.3.1)*

\# It is generally believed that the earth's core has a density of 13,000 kg/m³, which is unlikely because the coupling ratio (φ) shows us that the earth's core pressure is insufficient to raise its density to this extent. However, due to the relatively small mass of the iron core (1% of the earth's mass), a significant variation in its density will result in only a minor variation in the calculation result. So, for this exercise, we will set the core density to the popular (but unlikely) density of 13,000 kg/m³

Calculation Procedure

Newton's famous formula may be used to calculate the pressure anywhere within any body (or mass) with one minor modification:
$p = G.m_1.m_2 / A.r^2$
where:
A is the spherical surface area at radius 'r'
m_1 is the mass inside 'r'
m_2 is the mass outside 'r'

Divide the interior up into spherical layers of known (or assumed) radii and assign densities to each layer @ 'r'.

Then iterate through the calculation using the above formula, altering the radii and/or densities until the total mass and polar moment of inertia reflect the known values (input data 7 and 8 above)

CalQlata has created a calculator for core pressure that uses a variable density in each layer. This calculator was used to determine the properties of the internal structure of the earth (refer to Chapter 4.4): (http://calqlata.com/proddetail.asp?prod=00086).

3.5 The Atom

The following calculation procedure provides the formulas required to convert atomic energy to electro-magnetic energy, electricity and the specific heat capacity for any atom.
Refer to Chapter 6.5.2 for a verification of SHC.

There are a few important things to remember when carrying out these calculations:

1) The temperature we feel (and measure) is that emitted by the proton-electron pair(s) with its electron(s) in the inner-most electron shell (T_1). It is this temperature (only) that is used to define the [measured] specific heat capacity of an atom.

2) The specific heat capacity of an atom includes the kinetic energy of all of its electrons. The kinetic energy of the electron(s) in each shell will vary with its orbital radius.

3) All electrons possess the same electrical charge, which means that the spacing between each shell will be equal.

4) You start the calculation procedure by selecting the temperature of the atom; i.e. the temperature that you would feel or measure (T_1)

5) 'n' refers to the electron shell number; 1 to 46 for atomic numbers (Z) 1 to 92 (note: each shell can hold up to two electrons)

PHILOSOPHIÆ NATURALIS PRINCIPIA MATHEMATICA Revision IV

Input Data:

measured temperature: T_1

Atomic:

orbital radius of shell 1: $R_1 = X_R / T_1$
Properties in Shell 'n':
orbital radius: $R_N = R_1.n$
electron temperature: $T_n = X_R / R_n$
electron velocity: $v_n = \sqrt{[T_N / X]}$
kinetic energy of electron: $KE_n = \frac{1}{2}.m_e.v_n^2$
potential energy: $PE_n = -2.KE_n$

Electro-Magnetic:

properties of electro-magnetism radiated by [proton-]electron in shell 'n':
frequency: $f_N = v_n / 2\pi R_n$
wavelength: $\lambda_n = c / f_n$
amplitude: $A_n = R_n$
energy: $E_n = KE_n$
charge: $Q = e'$

Electrical:

current: $A_n = Q.f_n$
voltage: $V_n = E_n / Q$
resistance: $\Omega_n = V_n / A_n$
temperature: $T_n = 2.E_n / k_B.Y$ (will be same as T_N above)

Specific Heat Capacity (SHC):

SHC = $\Sigma KE_n / T_1.m.Y$ {J / kg.K}

Where:
m = atomic mass

3.5.1 Atomic Particles

The only two particles in the universe necessary to make it work are the proton and the electron. All the calculations in this section of the book concern these two particles alone.

Sub-atomic particles are not considered here as they are unnecessary.

3.5.1.1 The Electron

The mass of an electron: m_e = 9.1093897E-31 kg (constant)

The electrical charge of an electron never changes, irrespective of its situation:

e = -1.60217648753E-19 C (constant)

The relationship between temperature (T), velocity (v) and orbital radius (R) may be defined as follows:
$v^2 = T/X$ and $R = X_R/T$
where: X & X_R are heat transfer coefficients (refer to Chapter 6.6)

It is the electro-static *potential* energy that holds an orbiting electron to its proton. This energy is *always* twice the kinetic energy in the electron:
$KE = \frac{1}{2}.m_e.v^2$
$PE = 2.KE = 2.\frac{1}{2}.m_e.v^2 = m_e.v^2$

At the time of ejection from its proton-electron pair:

Angular velocity in an electron is: $\omega = 2\pi$ / orbital period
(at the time of ejection)

The linear velocity of an electron is: $v = \sqrt{[2.KE / m]}$
(at the time of ejection)

3.5.1.2 The Proton

The mass of a proton: m_p = 1.67262163783E-27 kg

The electrical charge of a lone proton, that is not part of proton-electron pair is the same as that of an electron:

e = -e = 1.60217648753E-19 C (constant)

After acquiring an orbiting electron, it will have the ability to top up its electrical charge to a maximum:

e' = -e.m_p/m_e = 2.94183820093364E-16 C

3.5.1.3 The Neutron

This is nature's energy storage facility; it is the third and last stage of universal energy production.

RAC = $k_B.R_i.Q_e$ = 96485.3317942156 C/mol (of electrons)
Note: Faraday's constant = 96485.3317942158 C/mol {exact}

Rest condition @ $\underline{T}$ = 1K: N_t = 1 N_v = 1.5 N_p = 2.5

RAM_p = $R_i.m_p/k_B$ = 1.00727638277235 N
Note: RAM_H = 1.00794 g/mol (hydrogen)
RAM_e = $RAM_p . m_e/m_p$ = 0.000548580318390698 g/mol
R_a = RAM_e / R_i = 15156.3563034308 J/g/K
R = $R_a.m_p$ = 1.38065156E-23 J/K
k_B = 1.38065156E-23 J/K
$k_B.N_A.L_n(N_t)$ = $c_p.L_n(\underline{T}).RAM_e$ = 3.371231032 J/K/mol
$\exp(2.5 \times L_n(\underline{T}))$ = 1

c_v = $N_t.R_a$ = 22734.5344551462 J/g/K
C_v = $m_e.c_v$ = 2.07097734E-23 J/K
c_p = $c_v + R_a$ = 37890.8907585769 J/g/K
C_p = $m_p.c_v$ = 3.4516289E-23 J/K

KE_e = $k_B.\underline{T}.N_p$ = 3.4516289E-23 J

X = $\underline{T}_n.t_n^2/(2\pi.R_n)^2$ = $\underline{T}_n/c^2$ = 6.9353271647894E-09 K.s²/m²
$\underline{T}$ = X.v² K

Or @ the speed of light; when a neutron is created:
$\underline{T}_n$ = X.c² = 623316124.71718 K

According to Newton's orbital motion formula; v = √[R.g]
'v' will be the speed of light (c)
when: R_n = orbital radius of 1.46677550700177 x ($r_p + r_e$)
R_n = 2.817937953839E-15m
g = $G.m_p / \varphi.R_n^2$ = 3.18940728807829E+31 m/s²

The magnetic field 'B' as described by Lorentz (R_n = 1E-07 . e/B), is actually 1/RC, where RC is the relative charge capacity of Quanta (1.75881869180547E+11 C/kg). And the orbital radius at which an electron and a proton combine to create a neutron may be calculated as follows: **R_n = µ'.e.RC = 2.817937953839E-15 m**

Where: 'e' is the elementary charge unit (1.6021764875E-19 C)
According to the heat-coefficient 'X', this orbital radius occurs when the electron is travelling at the speed of light (Table 4.5.3-4).

At the orbital radius; 'R_n', the attractive magnetic force exceeds the repulsive electrical force and the electron combines with its proton to create a neutron, which occurs when the electro-magnetic energy is equivalent to a temperature of 623316124.71718 K
Moreover, 'R_n' occurs when: KE = ½.m.c²

Note: In circular orbits; PE = -2.KE = -2 . ½.m.c² = -m.c²

The magnetic constant (μ_o), which controls this union between an electron and a proton, is referred to as;
μ_o = 1E-07 . 4π H/m,
but what exactly is 1E-07 and what is a *Henry*?
$m_e.R_n/e^2$ = 1E-07 (*exactly*) kg.m/C²
μ_o = 4.π.$R_n.m_e/e^2$ kg.m/C²
and; Henry = kg.m²/C²

Verifying that 'R_n' is a real and important physical constant (that I refer to as the **neutronic radius**) and that it occurs when PE = m.c²

This is the rationale behind E = m.c²

Moreover, all of the above can *only* be determined using Newton's laws of motion and Coulomb's electrical force. None of this is achievable with Relativity, in which mass is claimed to vary with velocity and gravity (magnetic charge) is claimed to deform the orbital path.

A neutron therefore has no electrical charge but possesses a magnetic charge and its magnitude is that of a proton plus that of an electron. It also stores the following energy:

E = KE − PE = mc² = -4.093555611312680E-14 J

It cannot be mere coincidence that Newton's and Coulomb's laws show that an electron orbiting its proton at the speed of light (c) comes within striking distance of each other, and that this occurs at a temperature of; ≈6.23E+08 K, or that Newton's and Coulomb's laws can explain all of the above clearly and accurately, but Relativity cannot.

3.5.2 Electron Shells

The innermost orbital radius is calculated thus; $R_1 = X_R/T$
then add this radius (R_1) to each subsequent shell radius (R_2 to R_n)

The electron temperature in subsequent shells can then be determined using the same formula: $T = X_R/R$

The orbital velocity of each electron is determined thus: $v = \sqrt{[T/X]}$

Their kinetic energies are: $KE = \frac{1}{2}.m.v^2$
Their potential energies are: $PE = -2.KE$

Electro-magnetic energy is equal to KE.

You use the formulas provided in Chapter 3.1.3 to find the properties of the electro-magnetic energy generated by each proton-electron pair.

Johannes Rydberg (refer to Chapter 3.1.3) gave us another useful constant (R_γ) that enables us to calculate electron shell number (n) from an electron's kinetic energy:

$n = (KE_n/R_\gamma) . (a_o/R_n) = (c/v_e)^2$

Where; 'v_e' is the velocity of the electron

3.5.3 Nucleus

Atomic lattice structure and density are defined thus; $\zeta = 3\sqrt{[\,m.n/\rho\,]} / R$
The resultant factors are listed below for Hydrogen to Uranium[4]:

Element	#	ζ	Element	#	ζ
Hydrogen	(1)	4.25	Silver	(47)	5.625
Helium	**(2)**	**1.5**	Cadmium	(48)	3.875
Lithium	(3)	19.5	Indium	(49)	6.125
Beryllium	(4)	16.25	Tin	(50)	6.875
Boron	(5)	17.5	Antimony	(51)	5.625
Carbon	(6)	18	Tellurium	(52)	4.75
Nitrogen	(7)	5.5	Iodine	(53)	3
Oxygen	(8)	5	**Xenon**	**(54)**	**2**
Fluorine	(9)	4.625	Caesium	(55)	6
Neon	**(10)**	**1.625**	Barium	(56)	6.875
Sodium	(11)	11.125	Lanthanum	(57)	8
Magnesium	(12)	10	Cerium	(58)	6.625
Aluminium	(13)	11.875	Praseodymium	(59)	7.875
Silicon	(14)	14	Neodymium	(60)	7.25
Phosphorus	(15)	6	Promethium	(61)	6.375
Sulphur	(16)	6.5	Samarium	(62)	5.5
Chlorine	(17)	6.5	Europium	(63)	5.875
Argon	**(18)**	**2.375**	Gadolinium	(64)	7
Potassium	(19)	10	Terbium	(65)	6.875
Calcium	(20)	10.75	Dysprosium	(66)	6.125
Scandium	(21)	12	Holmium	(67)	6.125
Titanium	(22)	10.375	Erbium	(68)	6.25
Vanadium	(23)	9.5	Thulium	(69)	5.25
Chromium	(24)	8	Ytterbium	(70)	4.625
Manganese	(25)	6.875	Lutetium	(71)	6.5
Iron	(26)	7.875	Hafnium	(72)	6.75
Cobalt	(27)	7.5	Tantalum	(73)	6.75
Nickel	(28)	7.5	Tungsten	(74)	6.5
Copper	(29)	7	Rhenium	(75)	6.375
Zinc	(30)	4.875	Osmium	(76)	5.875
Gallium	(31)	7.375	Iridium	(77)	5.5
Germanium	(32)	8.5	Platinum	(78)	5.25
Arsenic	(33)	4.375	Gold	(79)	4.75
Selenium	(34)	4.875	Mercury	(80)	2.375
Bromine	(35)	5.375	Thallium	(81)	4.125
Krypton	**(36)**	**2.25**	Lead	(82)	4.5
Rubidium	(37)	7	Bismuth	(83)	4.5
Strontium	(38)	7.625	Polonium	(84)	3.75
Yttrium	(39)	9.5	Astatine	(85)	3
Zirconium	(40)	9.5	**Radon**	**(86)**	**2**
Niobium	(41)	8.875	Francium	(87)	4.75
Molybdenum	(42)	8.25	Radium	(88)	5.625
Technetium	(43)	7.625	Actinium	(89)	6
Ruthenium	(44)	7.25	Thorium	(90)	7
Rhodium	(45)	6.875	Protactinium	(91)	5.75
Palladium	(46)	6.25	Uranium	(92)	5.5

Note; the lowest values for ζ occur at **the noble gases**.

PHILOSOPHIÆ NATURALIS PRINCIPIA MATHEMATICA Revision IV

Isotopic factors ($\Gamma = 9.[\psi-1]$) are listed below, for Hydrogen to Uranium[4]:

Element	№	Value	Element	№	Value
Hydrogen	(1)	0.07146	Silver	(47)	2.655613
Helium	**(2)**	**0.011709 (0)**	Cadmium	(48)	3.077063
Lithium	(3)	2.823	Indium	(49)	3.08902
Beryllium	(4)	2.2774095	Tin	(50)	3.3642
Boron	(5)	1.45998	Antimony	(51)	3.485294
Carbon	(6)	0.01605	Tellurium	(52)	4.084615
Nitrogen	(7)	0.008614286	Iodine	(53)	3.549816
Oxygen	(8)	0.0001125	**Xenon**	**(54)**	**3.984 (4)**
Fluorine	**(9)**	**0.998403 (1)**	Caesium	(55)	3.748016
Neon	(10)	0.16173	Barium	(56)	4.070411
Sodium	(11)	0.809811818	Lanthanum	(57)	3.932447
Magnesium	(12)	0.22875	Cerium	(58)	3.742138
Aluminium	(13)	0.679526308	Praseodymium	(59)	3.494387
Silicon	(14)	0.054964286	Neodymium	(60)	3.636
Phosphorus	(15)	0.5842566	Promethium	(61)	3.393443
Sulphur	(16)	0.0365625	Samarium	(62)	3.826452
Chlorine	(17)	0.769235294	Europium	(63)	3.709143
Argon	**(18)**	**1.974 (2)**	Gadolinium	(64)	4.113281
Potassium	(19)	0.520247368	Terbium	(65)	4.005047
Calcium	(20)	0.0351	Dysprosium	(66)	4.159091
Scandium	(21)	1.266818571	Holmium	(67)	4.154819
Titanium	(22)	1.581954545	Erbium	(68)	4.137221
Vanadium	(23)	1.933630435	Thulium	(69)	4.034897
Chromium	(24)	1.4985375	Ytterbium	(70)	4.248
Manganese	(25)	1.77769764	Lutetium	(71)	4.178916
Iron	(26)	1.331653846	Hafnium	(72)	4.31125
Cobalt	(27)	1.6444	Tantalum	(73)	4.308645
Nickel	(28)	0.865735714	Tungsten	(74)	4.360135
Copper	(29)	1.721172414	Rhenium	(75)	4.34484
Zinc	(30)	1.6155	Osmium	(76)	4.527237
Gallium	(31)	2.24216129	Iridium	(77)	4.466922
Germanium	(32)	2.4215625	Platinum	(78)	4.509
Arsenic	(33)	2.433163636	Gold	(79)	4.439227
Selenium	(34)	2.901176471	Mercury	(80)	4.566375
Bromine	(35)	2.546742857	Thallium	(81)	4.709256
Krypton	**(36)**	**2.9495 (3)**	Lead	(82)	4.741463
Rubidium	(37)	2.789464865	Bismuth	(83)	4.660523
Strontium	(38)	2.752105263	Polonium	(84)	4.390929
Yttrium	(39)	2.516734615	Astatine	(85)	4.233918
Zirconium	(40)	2.5254	**Radon**	**(86)**	**5.0244 (5)**
Niobium	(41)	2.394083415	Francium	(87)	5.071034
Molybdenum	(42)	2.558571429	Radium	(88)	5.116193
Technetium	(43)	2.701297674	Actinium	(89)	4.957888
Ruthenium	(44)	2.673409091	Thorium	(90)	5.20381
Rhodium	(45)	2.5811	Protactinium	(91)	5.142857
Palladium	(46)	2.821304348	Uranium	(92)	5.285437

Note; 'Γ = 0 to 5' occur at **the noble gases** (except Neon).

3.5.4 How They Work

An electron absorbs electro-magnetic energy and converts it into velocity. This means that at the 'speed-of-light' its kinetic energy reaches:
$KE = \frac{1}{2} \cdot m_e \cdot c^2$
and the coincident *potential* energy between the proton and its electron is:
$PE = m_e \cdot c^2$

There are three key energy conditions for the proton-electron pair according to Planck, each of which relate to electron velocities, shell radii and associated temperatures. These energies are listed below:

Neutronic: $KE_n = \frac{1}{2} \cdot m_e \cdot c^2 = 4.0935556113127E-14$ J

Mean: $KE_m = \frac{1}{2} \cdot m_e \cdot v_m^2 = 2.3771466644364E-17$ J

Minimum: $KE_o = \frac{1}{2} \cdot m_e \cdot v_o^2 = 1.3804200555196E-20$ J

There is also a fourth energy level that appears to be when an orbiting electron may leave its proton partner and continue in free-flight. This condition is referred to as the 'cold' energy level:

Cold: $KE_c = \frac{1}{2} \cdot m_e \cdot v_c^2 = 8.0161630672150E-24$ J

$\delta = KE_n / KE_c = 5.1066271693683E+09$

The potential energy between a proton and its orbiting electron is twice the kinetic energy of the electron.

The energy stored within each neutron is therefore: $-KE_n = PE_n + KE_n$

The orbiting electron defines the properties of the electro-magnetic energy emitted by the proton-electron pair (refer to Chapter 3.1.3)

3.5.4.1 Planck Electron Velocities

Neutronic: c = 299792459 m/s

Mean: $v_m = \sqrt{[T_m/X]}$ = 7224342.80705 m/s

Minimum: $v_o = \sqrt{[T_o/X]}$ = 174090.866621 m/s

Cold: $v_c = \sqrt{[T_c/X]}$ = 4195.2092599072 m/s

This means that an electron is unlikely to exceed (or even achieve) the *'speed of light'* in free-flight unless it is provided with artificial energy greater than ½.m.c² or knocked from its orbit, and most electrons in free-flight will be travelling at little more than 4000m/s

3.5.4.2 Planck Shell Radii

Neutronic: $R_n = X_R/\underline{T}_n = 2.81793795383896\text{E-}15$ m

Mean: $R_m = X_R/\underline{T}_m = 4.85261843362263\text{E-}12$ m

Minimum: $R_o = X_R/\underline{T}_o = 8.35643156381572\text{E-}09$ m

Cold: $R_c = X_R/\underline{T}_c = 1.43901585166681\text{E-}05$ m [2]

3.5.4.3 Planck Temperatures

Neutronic: $T_n = X \cdot c^2 = 623316124.71718$ K

Mean: $T_m = X \cdot v_m^2 = 361962.55467156$ K

Minimum: $T_o = X \cdot v_o^2 = 210.19332853584$ K

Cold: $T_c = X \cdot v_c^2 = 0.122060237421696$ K

The maximum possible natural temperature cannot exceed; T_n

4 Calculation Results

A collection of [mostly] tabulated calculation results for selected examples using the formulas provided in section 3 (above).

PHILOSOPHIÆ NATURALIS PRINCIPIA MATHEMATICA Revision IV

4.1 Energy

The calculation results for energy can be found Chapters 4.2 to 4.5 under the symbols PE & KE

4.1.1 Electricity

Refer to Chapter 6.9

4.1.2 Magnetism

Refer to Chapters 6.7 & 6.8

4.1.3 Electro-Magnetic Energy

Tables -1 to -2 below show the properties of electro-magnetic radiation emitted by a proton-electron pair at Planck's temperatures (refer to Chapter 3.5.4.3).

	minimum (T_o)	neutronic (T_n)
T (K)	210.19332853584	623316124.71718
f (Hz)	3.31570021944219E+12	1.69320448260835E+22
λ (m)	9.04160325599144E-05	1.77056263481051E-14
A (m)	8.3564315638157E-09	2.817937953839E-15 #
Spectra	Infra-red	γ

Table 4.1.3-1: Planck's Temperature Boundaries
T = temperature; f = frequency; λ = wavelength; A = amplitude
the neutronic radius (R_n) is 2.81793795383896E-15 m

	cold (T_c)	mean (T_m)
T_c (K)	0.122060237421696	361962.55467156
f (Hz)	46398953.1627279	2.3694215485103E+17
λ (m)	6.46119014686784	1.26525589837944E-09
A (m)	1.43901585166681E-05	4.85261843362268E-12
Spectra	radio	X

Table 4.1.3-2: Planck Temperatures
T = temperature; f = frequency; λ = wavelength; A = amplitude

We may deduce from the above Tables that:

The lowest possible electro-magnetic energy is that associated with a temperature of 0.12206 K and;

The highest possible electro-magnetic energy is that associated with a temperature of 623316124.71718 K

4.2 Orbits

The following Tables comprise the calculation results from Chapter 3.2 for the orbital systems with which we are most familiar:
Galactic: Milky Way
Solar: Solar System
Lunar: Those in our solar system
Atomic: Proton-electron pair

Tables -1: *Input Data*

G = Newton's gravitational constant
t = orbital period
R = radial distance between the centres of the force-centre and the satellite
^{P} = perigee & ^{A} = apogee
r = radius of m_2
m_2 = satellite mass

Tables -2: Orbital Shape

a & b = major and minor orbital semi-axes
e = orbital eccentricity
p = orbital half parameter
f = distance between orbital 'focus' and satellite (R^P)
x' = distance between orbital 'focus' and orbit centre ($a - f$)
A = total swept area of orbit
L = circumferential length of orbit
K = constant of proportionality
v = satellite curvilinear velocity
v_c = centrifugal velocity at 'θ'
g = gravitational acceleration between force-centre and satellite

---

Tables -3: Masses

m_1 = force-centre mass
m_2 = satellite mass

Tables -4: Orbital Performance

F = gravitational force between force-centre and satellite
PE = gravitational energy between force-centre and satellite
KE = kinetic energy in satellite
E = total energy should always be the same, irrespective of radial distance
h = constant of motion should always be the same, irrespective of radial distance

Suffixes: A = @ apogee; P = @ perigee; none = @ R (θ)

4.2.1 Galactic

The following Tables show the orbital properties of our sun based upon two galactic population options;
1) 100bn star-systems, which is the popularly held belief, but appears far too high to me, and;
2) 10bn star-systems, which also seems too high

The reason for performing both calculations is to show that irrespective of the number of star-systems in the Milky Way, the orbital calculations can be performed successfully without the need for dark matter, moreover the number of star-systems orbiting Hades has no effect on its orbital shapes or velocities.

You will notice that the results in both calculation options are exactly the same, despite the difference in star-system populations. This is because:

1) Hades' mass *only* defines the orbital radii and period, and
2) Hades' spin-rate is *only* defined by star-system population

It is difficult to understand therefore, how anybody could have declared 100 years ago that Newton's laws of orbital motion would predict the ejection of the Milky Way's stars into outer space through centrifugal force when:

3) Newton's laws of orbital motion are independent of star-system population
4) The presence of a galactic force-centre was not predicted at that time
5) The mathematical laws of planetary spin have only just been discovered

Note: All three values for total energy (E) should be identical if the input data is correct. The small difference at θ simply means that NASA's input data is slightly out.

Property	The Sun (10bn)	The Sun (100bn)	units
t	7.25825E+15	7.25825E+15	s
R^P	2.4653729E+20	2.4653729E+20	m
$R\ (\theta = 45°)$	2.533231129E+20	2.533231129E+20	m
R^A	2.5452510E+20	2.5452510E+20	m
m_2	1.9885E+30	1.9885E+30	kg

Table 4.2.1-1: *Input Data*

Property	The Sun (10bn)	The Sun (100bn)	units
a	2.505312E+20	2.505312E+20	m
e	0.015941744	0.015941744	
b	2.504994E+20	2.504994E+20	m
p	2.504675E+20	2.504675E+20	m
f	2.465373E+20	2.465373E+20	m
x'	3.993904E+18	3.993904E+18	m
A	1.971598E+41	1.971598E+41	m²
L	1.574034E+21	1.574034E+21	m
K	3.3502574E-30	3.3502574E-30	s²/m³
v^P	220360.56213	220360.56213	m/s
v	214457.71443	214457.71443	m/s
vc	215677.54520	215677.54520	m/s
v^A	213444.94588	213444.94588	m/s
g^P	-1.93872549E-10	-1.93872549E-10	m/s²
g	-1.83625048E-10	-1.83625048E-10	m/s²
g^A	-1.81894819E-10	-1.81894819E-10	m/s²

Table 4.2.1-2: Orbital Shape

Property	The Sun (10bn)	The Sun (100bn)	units
m_1	Hades: **1.76572E+41**	Hades: **1.76572E+41**	kg
m_2	1.9885E+30	1.9885E+30	kg

Table 4.2.1-3: Masses

Property	The Sun (10bn)	The Sun (100bn)	units
F^P	-3.8551556E+20	-3.8551556E+20	N
Fc^P	3.8551556E+20	3.8551556E+20	N
PE^P	-9.5043962E+40	-9.5043962E+40	J
KE^P	4.8279564E+40	4.8279564E+40	J
E^P	**-4.6764398E+40**	**-4.6764398E+40**	J
h^P	5.4327096E+25	5.4327096E+25	m²/s
F	-3.6513841E+20	-3.6513841E+20	N
Fc	3.6514104E+20	3.6514104E+20	N
PE	-9.2497998E+40	-9.2497998E+40	J
KE	4.5727657E+40	4.5727657E+40	J
E	**-4.6770342E+40**	**-4.6770342E+40**	J
h	5.4327096E+25	5.4327096E+25	m²/s
F^A	-3.6169785E+20	-3.6169785E+20	N
Fc^A	3.6160593E+20	3.6160593E+20	N
PE^A	-9.2061180E+40	-9.2061180E+40	J
KE^A	4.5296782E+40	4.5296782E+40	J
E^A	**-4.6764398E+40**	**-4.6764398E+40**	J
h^A	5.4327096E+25	5.4327096E+25	m²/s

Table 4.2.1-4: Orbital Performance

PHILOSOPHIÆ NATURALIS PRINCIPIA MATHEMATICA Revision IV

4.2.2 Solar

A solar system is simply a collection of planets and comets orbiting a star. Our solar system is known to us and therefore enables us to accurately identify the orbital properties of any and every member in great detail.

Whilst 21 solar orbits have been analysed in the creation of this book, I have chosen the results from just three to include here:

Tables -1 to -4 below show the orbital properties of Mercury, Earth and Jupiter, calculated using the formulas provided in Chapter 3.2.2

The 21 satellites analysed included; three of the largest asteroids (Ceres, Pallas & Vesta), three of its outermost planets (Eris, Haumea & MakeMake) and four of its comets (Halley's, C/2001 OG108 Loneos, 3200 Phaeton & P/2009 WX51 Catalina (CSS)). Whilst they have been analysed with similar accuracy to the planets, as their input data is suspect, their orbital output data is considered (by me) to be too unreliable to warrant inclusion here.

Note: All three values for total energy (E) should be identical if the input data is correct. The small difference at θ simply means that the input data is slightly out.

PHILOSOPHIÆ NATURALIS PRINCIPIA MATHEMATICA Revision IV

Property	Mercury	Earth	Jupiter	units
t	7600521.6	31558118.4	374335689.6	s
R^P	4.60012E+10	1.47095E+11	7.4052E+11	m
$R\,(\theta=45°)$	6.489479E+10	1.520941962E+11	8.036758283E+11	m
R^A	6.981450E+10	1.5209420E+11	8.156104331E+11	m
m_2	2439700	5.9645198E+24	1.89819E+27	kg

Table 4.2.2-1: *Input Data*

Property	Mercury	Earth	Jupiter	units
a	5.790785E+10	1.49595E+11	7.78065E+11	m
e	0.205613749	0.016709147	0.048254588	
b	5.667055E+10	1.49574E+11	7.77159E+11	m
p	5.545968E+10	1.49553E+11	7.76253E+11	m
f	4.60012E+10	1.47095E+11	7.4052E+11	m
x'	1.190665E+10	2499598078	37545216557	m
A	1.030967E+22	7.02945E+22	1.89966E+24	m²
L	3.599701E+11	9.39865E+11	4.88588E+12	m
K	2.974914E-19	2.9749144E-19	2.974914E-19	s²/m³
v^P	58974.212404	30286.0088	13705.90205	m/s
v	41804.350699	29436.3192	12628.84142	m/s
v_c	44699.545376	29611.76945	12844.06876	m/s
v^A	38858.468158	29290.5356	12444.04703	m/s
g^P	-0.062711465	-0.0061332328	-2.419979E-04	m/s²
g	-0.0315112486	-0.0057939183	-2.054582E-04	m/s²
g^A	-0.0272266369	-0.0057366715	-1.994894E-04	m/s²

Table 4.2.2-2: Orbital Shape

Property	Mercury	Earth	Jupiter	units
m_1	1.9885E+30	1.9885E+30	1.9885E+30	kg
m_2	3.3011E+23	5.96451976771E+24	1.89819E+27	kg

Table 4.2.2-3: Masses

Property	Mercury	Earth	Jupiter	units
F^P	-2.0701682E+22	-3.6581678E+22	-4.589200E+23	N
Fc^P	2.070168E+22	3.6581788E+22	4.593581E+23	N
PE^P	-9.523022E+32	-5.3809820E+33	-3.398394E+35	J
KE^P	5.740543E+32	2.7354550E+33	1.782892E+35	J
E^P	**-3.782479E+32**	**-2.6455270E+33**	**-1.615503E+35**	J
h^P	2.712885E+15	4.4549138E+15	1.014465E+16	m²/s
F	-1.040218E+22	-3.4557837E+22	-3.896267E+23	N
F_c	1.016378E+22	3.4557932E+22	3.896404E+23	N
PE	-6.750472E+32	-5.2300158E+33	-3.131336E+35	J
KE	2.884507E+32	2.5841189E+33	1.513689E+35	J
E	**-3.865964E+32**	**-2.6458968E+33**	**-1.617647E+35**	J
h	2.712885E+15	4.4549138E+15	1.014465E+16	m²/s
F^A	-8.987785E+21	-3.4216388E+22	-3.783076E+23	N
Fc^A	8.607808E+21	3.4206938E+22	3.777870E+23	N
PE^A	-6.274777E+32	-5.2041141E+33	-3.085516E+35	J
KE^A	2.492298E+32	2.5585865E+33	1.469714E+35	J
E^A	**-3.782479E+32**	**-2.6455275E+33**	**-1.615802E+35**	J
h^A	2.712885E+15	4.4549138E+15	1.014465E+16	m²/s

Table 4.2.2-4: Orbital Performance

4.2.3 Lunar

A lunar system is simply a collection of moons orbiting a planet, just as planets orbit their stars. There is no limit to the number of moons any given planet can host. Whilst our own planet (Earth) can claim only one, Jupiter and Saturn host upwards of 100 between them. In fact, there are probably over 200 moons in our solar system, 152 of which have been analysed in the creation of this book.

Tables -1 to -4 below show the orbital properties of three of the best-known; our own moon, Phobos and Titan, each of which were calculated using the formulas provided in Chapter 4

Whilst it is by no means definitive, moons orbiting closer to their planet tend to orbit in a prograde direction and those further away tend to orbit in a retrograde direction.

The orbital planes of the moons in our solar system indicates that they have most probably come from outside our solar system. Jupiter's lunar orbital planes are only 3° from Jupiter's own orbital plane because of the strong influence of spin between the solar system's two most massive bodies.

It is most likely that planets orbiting closest the solar force-centre will not collect moons, because the sun's gravitational energy along with its proximity will be sufficient to trap or deflect most, if not all, of them from these planets. This is why Mercury and Venus are the only planets in our solar system without moons, and why our Earth has only one, whilst much smaller planets such as Mars and Pluto have managed to trap many more.

Note: All three values for total energy (E) should be identical if the input data is correct. The small difference at θ simply means that the input data is slightly out.

PHILOSOPHIÆ NATURALIS PRINCIPIA MATHEMATICA Revision IV

Property	Moon	Phobos	Titan	units
t	2360620.8	27553.84387	1378079.998	s
R^P	359508000	9230818.576	1.186590E+09	m
R ($\theta = 45°$)	398868087.4	9473266.673	1.246426E+09	m
R^A	406504000	9516150	1.257304E+09	m
m_2	1737494.514	11166.66667	2574707.348	kg

Table 4.2.3-1: *Input Data*

Property	Moon	Phobos	Titan	units
a	383006000	9373484.288	1221946914	m
e	0.061351519	0.015220137	0.0289348727	
b	382284501.4	9372398.529	1221435284.29	m
p	381564362	9371312.896	1220923868	m
f	359508000	9230818.576	1186590036	m
x'	23498000	142665.712	35356878.42	m
A	4.599834E+17	2.759953E+14	4.688918E+18	m^2
L	2.404232E+09	5.889193E+07	7.676112E+09	m
K	9.918265E-14	9.218520E-13	1.040859E-15	s^2/m^3
v^P	1084.020134	2170.247491	5734.922357	m/s
v	977.0496126	2114.704625	5459.6124	m/s
v_c	998.1341195	2126.191187	5515.7452	m/s
v^A	958.6963727	2105.174976	5412.376679	m/s
g^P	-0.0030796892	-0.5025949109	-0.026938071	m/s^2
g	-0.0025018740	-0.4771984141	-0.024413781	m/s^2
g^A	-0.0024087646	-0.4729072361	-0.02399316	m/s^2

Table 4.2.3-2: **Orbital Shape**

Property	Moon	Phobos	Titan	units
m_1	5.964367E+24	6.417101E+23	5.683400E+26	kg
m_2	7.346377E+22	1.065853E+16	1.345525E+23	kg

Table 4.2.3-3: **Masses**

Property	Moon	Phobos	Titan	units
F^P	-2.2624559E+20	-5.3569229E+15	-3.624585E+21	N
Fc^P	2.2624559E+20	5.3569229E+15	3.624585E+21	N
PE^P	-8.1337101E+28	-4.9448783E+22	-4.300896E+30	J
KE^P	4.3163628E+28	2.5100700E+22	2.212671E+30	J
E^P	**-3.8173473E+28**	**-2.4348083E+22**	**-2.088225E+30**	J
h^P	3.8971391E+11	2.0033161E+10	6.805002E+12	m^2/s
F	-1.8379710E+20	-5.0862336E+15	-3.284935E+21	N
F_c	1.8349395E+20	5.0863023E+15	3.284231E+21	N
PE	-7.3310799E+28	-4.8183247E+22	-4.094428E+30	J
KE	3.5065212E+28	2.3832343E+22	2.005328E+30	J
E	**-3.8245587E+28**	**-2.4350904E+22**	**-2.089100E+30**	J
h	3.8971391E+11	2.0033161E+10	6.805002E+12	m^2/s
F^A	-1.7695694E+20	-5.0404959E+15	-3.228340E+21	N
Fc^A	1.7629087E+20	5.0393283E+15	3.225637E+21	N
PE^A	-7.1933704E+28	-4.7966115E+22	-4.059004E+30	J
KE^A	3.3760231E+28	2.3618032E+22	1.970778E+30	J
E^A	**-3.8173473E+28**	**-2.4348083E+22**	**-2.088225E+30**	J
h^A	3.8971391E+11	2.0033161E+10	6.805002E+12	m^2/s

Table 4.2.3-4: **Orbital Performance**

4.2.4 Atomic

All electrons in an atom must obey Newton's laws of orbital motion, and their spacing between adjacent electrons in the same and adjacent shells is maintained by Coulomb's force-law. These conditions define the amount of electro-magnetic energy any given electron is able to absorb.

Each shell can hold up to two identical electrons, both of which will absorb the same amount of electro-magnetic energy (heat or temperature) from their surroundings.

This means that the electro-magnetic energy (heat or temperature) absorbed in each shell will be different from each of the other shells. Moreover, in accordance with Newton's laws of motion, electron(s) in the innermost shell will absorb the most energy and the outermost shell will absorb the least.

The heat we feel from matter is the sum of the electro-magnetic energy radiated by each proton-electron pairing within an atom. Its temperature is that generated by the proton-electron pairs in the innermost shell.

The Tables in this Chapter list electron orbital performance at the maximum possible temperature; when the orbiting electron achieves *light-speed*.

Below is listed the descriptions of the additional symbols used in the *electron* Tables.

Tables -1 to -4: (specific to the atom)

T = the temperature of the electron
X = velocity heat coefficient [5]
X_R = radial heat coefficient [5]

Tables -1 to -4 below show the orbital properties of a proton-electron pair at the maximum possible temperature (when the neutron is created).

Sym	Newton	Coulomb	units
T	623316124.7171790	623316124.7171790	K
m_2	9.1093897E-31	9.1093897E-31	kg
X		6.9353271647894E-09	$K.s^2/m^2$
X_R		1.75646616508036E-06	$K.m$

Table 4.2.4-1: *Input Data* ($\underline{T}_n$)

Sym	Newton	Coulomb	units
R	2.817937953839E-15	2.817937953839E-15	m
d	8.852813174052E-15	8.852813174052E-15	m
ℓ	2.817937953839E-15	2.817937953839E-15	m
a	2.817937953839E-15	2.817937953839E-15	m
e	0	0	
b	2.817937953839E-15	2.817937953839E-15	m
p	2.817937953839E-15	2.817937953839E-15	m
f	2.817937953839E-15	2.817937953839E-15	m
x'	0	0	m
A	2.494667824141E-29	2.494667824141E-29	m^2
L	1.770562634810E-14	1.770562634810E-14	m
K	0.15587874533403	0.15587874533403	s^2/m^3

Table 4.2.4-2: *Orbital Shape* ($\underline{T}_n$)

Sym	Newton	Coulomb	units
m_1	1.67262163783E-27	1.6726216378300E-27	kg
m_2	9.1093897E-31	9.1093897E-31	kg

Table 4.2.4-3: *Masses* ($\underline{T}_n$)

Sym	Newton	Coulomb	units
v	6.2938005855237E-12	299792459	m/s
g	1.4057061035135E-08	3.189407288078E+31	m/s^2
F	1.2805124700573E-38	29.05355389912620	N
F_c	1.2805124700573E-38	29.05355389912620	N
PE	-3.6084046897386E-53	-8.1871112226254E-14	J
KE	1.8042023448693E-53	4.0935556113127E-14	J
E	-1.8042023448693E-53	-4.0935556113127E-14	J
h	1.7735539543841E-26	8.4479654849081E-07	m^2/s
PE/KE	-2	-2	

Table 4.2.4-4: *Orbital Performance* ($\underline{T}_n$)

4.3 Spin

Below is listed the descriptions of the symbols used in Chapters 4.3.4 to 4.3.6.

θ = planetary tilt
ψ = Sign(Cos(θ))
Δ = radial modifier for J (density variable)
J = polar moment of inertia
R_{ave} = average orbital radius
ω_0 = rotational velocity due to orbit
E_0 = natural rotational energy due to orbit
E_1 = rotational energy induced by its force-centre
E_2 = total rotational energy
E_3 = rotational energy induced by its sub-satellites (e.g. moons)
ω_2 = total rotational velocity

4.3.1 Polar Moment of Inertia

Below is listed the polar moment of inertia and radial modifier for each of the planets in our solar system:

Planet	J {kg.m²}	Δ
Our Sun	3.90008074E+46	0.318
Mercury	5.19308435E+35	0.813
Venus	3.30912713E+37	0.681
Earth	1.08209548E+37	0.334
Mars	1.58326892E+31	0.00232
Jupiter	1.92585538E+39	0.0228
Saturn	1.52392272E+38	0.0141
Uranus	1.38906233E+37	0.0249
Neptune	1.05696850E+38	0.0652
Pluto	5.48499917E+35	8.642

4.3.2 The Earth's Core

$\delta\omega = \omega + \omega_m =$ **6.95342605725E-05 ᶜ/s**

I.e. the earth's mantle is rotating 6.9534E-05 radians per second faster than its inner core, which is responsible for creating the earth's internal heat (through friction) and its magnetic field.

The positive value for '$\delta\omega$' shows the correct rotational direction according to the right-hand-rule (North is currently at the top of planet).

4.3.3 The Earth's Magnetic Field

The angular tilt (β) between the two axes can be calculated thus:
$\beta = \text{sign}(\omega/\omega_m) \cdot \frac{1}{2} \cdot \sqrt{[\ |A\sin(\omega/\omega_m)|\ ]}$
  = 0.109553685228394 radians (**6.27696379369167°**)

4.3.4 Our Sun

The following Table lists the spin characteristics of our sun based upon two different galactic populations. Note: they are identical.

Sym	Sun (100bn)	Sun (10bn)	units
θ	45	45	°
ψ	1	1	
Δ	0.318284697814735	0.318284697814735	
J	3.90008074E+46	3.90008074E+46	kg.m²
R_{ave}	2.50531194E+20	2.50531194E+20	m
ω_0	8.65661425E-16	8.65661425E-16	°/s
E_0	1.46130117E+16	1.46130117E+16	J
E_1	2.30014080E+16	2.30014080E+16	J
E_2	1.60100474E+35	1.60100474E+35	J
E_3	-1.60100474E+35	-1.60100474E+35	J
ω_2	**2.86532908E-06**	**2.86532908E-06**	°/s

Table 4.3.4-1: Spin Velocity of Our Sun (star-systems)

4.3.5 Our Planets

Planetary spin has been analysed for all the planets within our solar system and is listed in Tables -1 to -3 below.

Sym	Mercury	Venus	Earth	units
θ	0.01	177.4	23.4	°
ψ	1	-1	1	
Δ	0.812862196423	0.68123190998	0.3342776983	
J	5.19308435E+35	3.30912713E+37	1.08209548E+37	kg.m²
R_{ave}	5.79078502E+10	1.08205784E+11	1.49594598E+11	m
ω_0	8.26678173E-07	3.23643522E-07	1.99098857E-07	°/s
E_0	1.77446862E+23	1.73307475E+24	2.14473245E+23	J
E_1	5.76562924E+23	2.51533292E+23	3.20799628E+23	J
E_2	3.99116062E+23	-1.48154146E+24	2.87701826E+28	J
E_3	0	0	-2.87700762E+28	J
ω_2	**1.23980080E-06**	**-2.99236920E-07**	**7.29211510E-05**	°/s

Table 4.3.5-1: Spin Velocity in the Inner Planets

Sym	Mars	Jupiter	Saturn	units
θ	25.2	3.1	26.7	°
ψ	1	1	1	
Δ	**0.0023170868178**	0.022780669614	0.014060010927	
J	1.58326892E+31	1.92585538E+39	1.52392272E+38	kg.m²
R_{ave}	2.27934435E+11	7.78065217E+11	1.42682770E+12	m
ω_0	1.05858984E-07	1.67848952E-08	6.75904500E-09	°/s
E_0	8.87115436E+16	2.71288224E+23	3.48099680E+21	J
E_1	1.55613559E+22	2.52841886E+26	9.19174032E+24	J
E_2	3.97741612E+22	2.97776807E+31	2.04408655E+30	J
E_3	-2.42128940E+22	-2.97774281E+31	-2.04407737E+30	J
ω_2	**7.08823600E-05**	**1.75852520E-04**	**1.63788410E-04**	°/s

Table 4.3.5-2: Spin Velocity in the Middle Planets

Sym	Uranus	Neptune	Pluto	units
θ	97.8	28.3	122.5	°
ψ	-1	1	-1	
Δ	0.024937619324	0.06523792741	8.64241984998	
J	1.38906233E+37	1.05696850E+38	5.48499917E+35	kg.m²
R_{ave}	2.86949539E+12	4.49637396E+12	5.90394017E+12	m
ω_0	2.36992355E-09	1.20822828E-09	8.03026194E-10	°/s
E_0	3.90086044E+19	7.71489543E+19	1.76850379E+17	J
E_1	2.80792156E+22	2.09359675E+21	6.27106256E+15	J
E_2	-7.1182956E+28	6.20291286E+29	-3.55514898E+25	J
E_3	7.11829844E+28	-6.20291284E+29	3.55514896E+25	J
ω_2	**-1.0123767E-04**	**1.0833825E-04**	**-1.1385592E-05**	°/s

Table 4.3.5-3: Spin Velocity in the Outer Planets

4.3.6 Hades

The following Table lists the spin characteristics of Hades based upon two different galactic populations; note: they are different.

Sym	Hades (100bn)	Hades (10bn)	units
θ	0	0	°
ψ	1	1	
Δ	1	1	
J	2.8141903980768E+65	2.8141903980768E+65	kg.m²
R_{ave}	0	0	m
ω_0	0	0	c/s
E_0	0	0	J
E_1	0	0	J
E_2	4.378161564734E+51	4.3781615647344E+50	J
E_3	-4.378161564734E+51	-4.3781615647344E+50	J
ω_2	**1.76394139222867E-07**	**5.57807245849103E-08**	c/s

Table 4.3.6-1: Spin Velocity of Hades (star-systems)

PHILOSOPHIÆ NATURALIS PRINCIPIA MATHEMATICA Revision IV

4.4 Core Pressure

A core pressure calculation using the above theory (refer to Chapter 3.4) for the earth is presented in Fig 33 in which it can be seen that the upper mantle material has a density of 1105kg/m³.

A description of the input and output data from CalQlata's program (*Cores*) for pressures within for our earth is listed below:

R_0 to R_6 = outside radii for each spherical layer
$ρ_0$ to $ρ_6$ = density at each spherical radius
m = mass of the planet
J = polar moment of area of the planet
p_6 = atmospheric pressure at the planet's outer surface
G = Newton's gravitational constant
Fm = mass-factor (must equal 1)
FJ = polar moment of area-factor (must equal 1)
p_0 to p_5 = pressure at each spherical radius (R_0 to R_6)

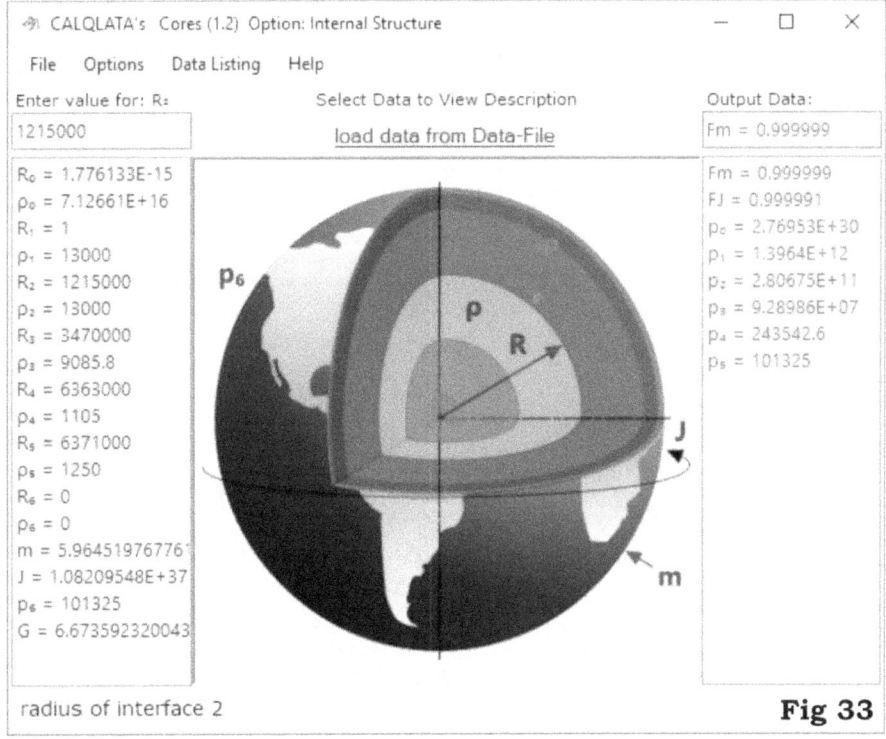

Fig 33

With kind permission of CalQlata

PHILOSOPHIÆ NATURALIS PRINCIPIA MATHEMATICA Revision IV

4.5 The atom

The following lists the symbol descriptions for Tables -1 to -4:

Tables -1: *Input Data*

T = electron temperature
m_2 = electron mass
X = Heat coefficient for electron velocity and temperature
X_R = Heat coefficient for electron orbital radius and temperature

---

Tables -2: Orbital Shape

R = radial distance between the centres of the force-centre and the satellite
d = arc distance between equi-spaced points on the surface of a sphere
ℓ = straight-line distance between equi-spaced points on the surface of a sphere
a & b = major and minor orbital semi-axes
e = orbital eccentricity
p = orbital half parameter
f = distance between orbital 'focus' and satellite (R^P)
x' = distance between orbital 'focus' and orbit centre ($a - f$)
A = total swept area of orbit
L = circumferential length of orbit
K = constant of proportionality

Tables -3: Masses

m_1 = force-centre mass
m_2 = satellite mass

Tables -4: Orbital Performance

v = satellite curvilinear velocity
g = gravitational acceleration between force-centre and satellite
F = gravitational force between force-centre and satellite
F_c = centrifugal force on electron
PE = gravitational energy between force-centre and satellite
KE = kinetic energy in satellite
E = total energy should always be the same, irrespective of radial distance
h = constant of motion should always be the same, irrespective of radial distance
PE/KE = confirmation of 2:1 relationship in circular orbits

4.5.1 Cold (T_c)

Tables -1 to -4 below show the orbital properties of a proton-electron pair at the minimum temperature.

Property	Newton	Coulomb	units
T	0.122060237421696	0.122060237421696	K
m_2	9.1093897E-31	9.1093897E-31	kg
X		6.9353271647894E-09	$K.s^2/m^2$
X_R		1.75646616508036E-06	$K.m$

Table 4.5.1-1: *Input Data* (T_c)

Property	Newton	Coulomb	units
R	1.439015851667E-05	1.439015851667E-05	m
d	1.439015851667E-05	1.439015851667E-05	m
ℓ	4.520801627996E-05	4.520801627996E-05	m
a	1.439015851667E-05	1.439015851667E-05	m
e	0	0	
b	1.439015851667E-05	1.439015851667E-05	m
p	1.439015851667E-05	1.439015851667E-05	m
f	1.439015851667E-05	1.439015851667E-05	m
x'	0	0	m
A	6.505505204927E-10	6.505505204927E-10	m^2
L	9.041603255991450E-05	9.041603255991450E-05	m
K	0.1558787453340300	0.1558787453340300	s^2/m^3

Table 4.5.1-2: **Orbital Shape** (T_c)

Property	Newton	Coulomb	units
m_1	1.67262163783E-27	1.6726216378300E-27	kg
m_2	9.1093897E-31	9.1093897E-31	kg

Table 4.5.1-3: **Masses** (T_c)

Property	Newton	Coulomb	units
v	8.8073631286363E-17	4195.2092599071500	m/s
g	5.3904649618566E-28	1.223042867389E+12	m/s^2
F	4.9103846001748E-58	1.1141174098854E-18	N
Fc	4.9103846001748E-58	1.1141174098854E-18	N
PE	-7.0661212774321E-63	-1.6032326134430E-23	J
KE	3.5330606387160E-63	8.0161630672150E-24	J
E	-3.5330606387160E-63	-8.0161630672150E-24	J
h	1.2673935153494E-21	6.0369726260658E-02	m^2/s
PE/KE	-2	-2	

Table 4.5.1-4: **Orbital Performance** (T_c)

4.5.2 Planck Minimum (T_o)

Tables -1 to -4 below show the orbital properties of a proton-electron pair at Planck's minimum temperature [3].

Property	Newton	Coulomb	units
T	210.19332853584	210.19332853584	K
m_2	9.1093897E-31	9.1093897E-31	kg
X		6.9353271647894E-09	$K.s^2/m^2$
X_R		1.75646616508036E-06	$K.m$

Table 4.5.2-1: *Input Data* (T_o)

Property	Newton	Coulomb	units
R	8.356431563816E-09	8.356431563816E-09	m
d	2.625250401111E-08	2.625250401111E-08	m
ℓ	8.356431563816E-09	8.356431563816E-09	m
a	8.356431563816E-09	8.356431563816E-09	m
e	0	0	
b	8.356431563816E-09	8.356431563816E-09	m
p	8.356431563816E-09	8.356431563816E-09	m
f	8.356431563816E-09	8.356431563816E-09	m
x'	0	0	m
A	7.396196699443E-10	7.396196699443E-10	m²
L	9.640713088869E-05	9.640713088869E-05	m
K	0.15587874533403	0.15587874533403	s²/m³

Table 4.5.2-2: Orbital Shape (T_o)

Property	Newton	Coulomb	units
m_1	1.67262163783E-27	1.6726216378300E-27	kg
m_2	9.1093897E-31	9.1093897E-31	kg

Table 4.5.2-3: Masses (T_o)

Property	Newton	Coulomb	units
v	3.6548390907795E-15	174090.86662108400	m/s
g	1.5985111201450E-21	3.6268626876711E+18	m/s²
F	1.4561460733184E-51	3.3038505610385E-12	N
Fc	1.4561460733184E-51	3.3038505610385E-12	N
PE	-1.2168185008604E-59	-2.7608401110393E-20	J
KE	6.0840925043021E-60	1.3804200555196E-20	J
E	-6.0840925043021E-60	-1.3804200555196E-20	J
h	3.0541412738857E-23	1.4547784128045E-03	m²/s
PE/KE	-2	-2	

Table 4.5.2-4: Orbital Performance (T_o)

4.5.3 Planck Mean (T_m)

Tables -1 to -4 below show the orbital properties of a proton-electron pair at Planck's mean temperature.

Property	Newton	Coulomb	units
T	361962.55467156	361962.55467156	K
m_2	9.1093897E-31	9.1093897E-31	kg
X		6.9353271647894E-09	$K.s^2/m^2$
X_R		1.75646616508036E-06	$K.m$

Table 4.5.3-1: Input Data (T_m)

Property	Newton	Coulomb	units
R	4.852618433623E-12	4.852618433623E-12	m
d	1.524495042174E-11	1.524495042174E-11	m
ℓ	4.852618433623E-12	4.852618433623E-12	m
a	4.852618433623E-12	4.852618433623E-12	m
e	0	0	
b	4.852618433623E-12	4.852618433623E-12	m
p	4.852618433623E-12	4.852618433623E-12	m
f	4.852618433623E-12	4.852618433623E-12	m
x'	0	0	m
A	2.193772531476E-16	2.193772531476E-16	m²
L	5.250500802222E-08	5.250500802222E-08	m
K	0.15587874533403	0.15587874533403	s²/m³

Table 4.5.3-2: Orbital Shape (T_m)

Property	Newton	Coulomb	units
m_1	1.67262163783E-27	1.6726216378300E-27	kg
m_2	9.1093897E-31	9.1093897E-31	kg

Table 4.5.3-3: Masses (T_m)

Property	Newton	Coulomb	units
v	1.5166683358448E-13	7224342.80705005	m/s
g	.7402920143405E-15	1.0755250944965E+25	m/s²
F	4.3181167250426E-45	9.7973772178982E-06	N
Fc	4.3181167250426E-45	9.7973772178982E-06	N
PE	-2.0954172818476E-56	-4.7542933288727E-17	J
KE	1.0477086409238E-56	2.3771466644364E-17	J
E	-1.0477086409238E-56	-2.3771466644364E-17	J
h	7.3598127242123E-25	3.5056979076300E-05	m²/s
PE/KE	-2	-2	

Table 4.5.3-4: Orbital Performance (T_m)

4.5.4 Neutron (T_n)

Tables -1 to -4 below show the orbital properties of a proton-electron pair at the maximum possible temperature (when the neutron is created).

Property	Newton	Coulomb	units
T	623316124.71718	623316124.71718	K
m_2	9.1093897E-31	9.1093897E-31	kg
X		6.9353271647894E-09	K.s²/m²
X_R		1.75646616508036E-06	K.m

Table 4.5.4-1: *Input Data* (T_n)

Property	Newton	Coulomb	units
R	2.817937953839E-15	2.817937953839E-15	m
d	8.852813174052E-15	8.852813174052E-15	m
ℓ	2.817937953839E-15	2.817937953839E-15	m
a	2.817937953839E-15	2.817937953839E-15	m
e	0	0	
b	2.817937953839E-15	2.817937953839E-15	m
p	2.817937953839E-15	2.817937953839E-15	m
f	2.817937953839E-15	2.817937953839E-15	m
x'	0	0	m
A	2.494667824141E-29	2.494667824141E-29	m²
L	1.770562634810E-14	1.770562634810E-14	m
K	0.15587874533403	0.15587874533403	s²/m³

Table 4.5.4-2: **Orbital Shape** (T_n)

Property	Newton	Coulomb	units
m_1	1.67262163783E-27	1.6726216378300E-27	kg
m_2	9.1093897E-31	9.1093897E-31	kg

Table 4.5.4-3: **Masses** (T_n)

Property	Newton	Coulomb	units
v	6.2938005855237E-12	299792459	m/s
g	1.4057061035135E-08	3.189407288078E+31	m/s²
F	1.2805124700573E-38	29.05355389912620	N
Fc	1.2805124700573E-38	29.05355389912620	N
PE	-3.6084046897386E-53	-8.1871112226254E-14	J
KE	1.8042023448693E-53	4.0935556113127E-14	J
E	-1.8042023448693E-53	-4.0935556113127E-14	J
h	1.7735539543841E-26	8.4479654849081E-07	m²/s
PE/KE	-2	-2	

Table 4.5.4-4: **Orbital Performance** (T_n)

223

PHILOSOPHIÆ NATURALIS PRINCIPIA MATHEMATICA Revision IV

The following Table is a check-list of the ratios between various Newton (^{N}) and Coulomb (^{C}) calculated properties:

Property Ratio		Value
$v^N : v^C$	$\sqrt{\varphi}$	2.0993858906650E-20
$g^N : g^C$	φ	4.40742111792335E-40
$F^N : F^C$	φ	4.40742111792335E-40
$Fc^N : Fc^C$	φ	4.40742111792335E-40
$PE^N : PE^C$	φ	4.40742111792335E-40
$KE^N : KE^C$	φ	4.40742111792335E-40
$E^N : E^C$	φ	4.40742111792335E-40
$h^N : h^C$	$\sqrt{\varphi}$	2.09938589066502E-20

All of which are either the coupling ratio or its square-root

5 The Physical Constants

All the physical constants (including electrical properties such as Volts, Amps, Henries, Farads, Ohms, etc.) are provided (to ≤15 decimal places) in terms of the same four basic units; length, time, mass and charge and two ratios: m_e, e, R_n, t_n & ξ_v, ξ_m

PHILOSOPHIÆ NATURALIS PRINCIPIA MATHEMATICA Revision IV

5.1 Introduction

Since venturing into this mathematical field almost three years ago, what struck me most, was the inability to define mechanical, electrical and magnetic properties in terms of the same units. It is inconvenient to define mechanical properties in terms of electrical and/or magnetic energy.

Formulas for permittivity, magnetic constant, etc. are far too obscure and Planck's universal energy constant (h) actually has incorrect units. The magnetic field constant (B) is simply the reciprocal of the relative charge capacity (RC) and therefore becomes redundant. Whilst heat capacity coefficients exist, there are no such coefficients for charge. Etc.

Every document I have used in the past offers approximate values for all constants. Some of the more overt publications actually add ±Tolerance values in brackets to show how clever they are! In fact, there are only a few (7) primary constants, the values for which are known accurately. As all the others can be calculated, it is a simple matter to establish accurate values for *all* constants. We shouldn't need to qualify them with '(±??)'. For this reason, I have provided values for all constants accurate to 15 significant figures, except where absolute accuracy requires less. I leave it up to you to round them off if preferred. Approximations are unnecessary.

In the light of my discoveries, not least the neutronic radius (R_n), I have defined the properties of *all* astronomic and atomic forces and energies in these same universal units, making it much easier to understand the connection between the orbital systems.

I do not claim that everything in this publication is true and exact, merely that it represents the best and most useable collection of natural constants compiled to date. Please let me know of any errors or omission, and I shall update this publication accordingly.

PHILOSOPHIÆ NATURALIS PRINCIPIA MATHEMATICA Revision IV

The physical constants are the most important part of evaluating and defining natural laws, all of which are calculated here from just five fundamental constants (mass, charge, length, time & temperature) and two ratios (mass and velocity)

Because everything in the universe is energy, and we have (to date) concentrated on defining nature in terms of mechanics, the true nature of our universe has become very difficult to reconcile.

For example; we cannot readily explain Volts, Amps, Ohms, etc. in terms of energy. Therefore, I have reduced everything to the same basic [metric] units (Imperial conversions are in parenthesis):

Mass: kilogram (kg)
(1kg = 2.20462262lb)

Length: metre (m)
(1m = 39.3700787401575in)

Time: second (s)
(1s = 1s)

Electricity: Coulomb (C)
(1C = 1C)

Temperature: Kelvin (K)
(1K = 1.8 R)

These are the fundamental units that define all others. *Everything* can be explained and described (mathematically) using them.
Temperature is only a form of measurement for electron kinetic energy; it is not required to explain any natural property.
The only constant not fully resolved is the unit of magnetic charge, which is currently described as mass (kg).

Notes:
1) Joules and Newtons remain useful, but they are merely compilations of the above.
2) Converting to imperial units ...
... between numerators or denominators: multiply by the conversion factor above
... across numerators and denominators: divide by the conversion factor above

5.2 Symbols

The following is an alphabetical list of the symbols explained in the Tables (in this Chapter 5) indicated; **Table 5.?**.

Those that are new, i.e. not currently known, are highlighted in **bold text**

Symbol	Description	Table(s)
a_o	Rydberg radius (also known as Bohr Radius)	4
A	electrical current	8
c	speed of electro-magnetic radiation	4
$C_?$	specific heat capacity	6
$C_?$	heat capacity	5, 6
e	elementary charge unit	3
e	natural logarithm	4
E	energy	6.10.1 & 2, 6.10
F	Farad	8
F	Force	6.10.1 & 2, 6.10
G	Newton's gravitational constant	4
h	Planck's constant	4
ℏ	Planck's constant (Dirac version)	4
h'	modified Planck's constant	4
H	Henry	8
k	Coulomb's constant	4
k'	Coulomb's constant (modified)	4
k_B	Boltzmann's constant	5
K	Constant of proportionality	Chapter 6.11.14
m	mass	6.10.1 & 2, 6.10
m_e	mass of an electron	3
m_p	mass of a proton	4
m_n	mass of a neutron	4
$N_?$	microstate	5
N_A	Avogadro's number	6.11.12
q	specific charge capacity	7
Q	charge capacity	5, 7

Table 5.1a

Symbol	Description	Table(s)
r	particle (or body) radius	6.10.1 & 2
R	orbital radius	4.5.1 to 4
$R_?$	gas constant	5
R_a	specific gas constant	6
RAC	relative atomic charge	7
RAM	relative atomic mass	6
$\mathbf{R_c}$	charge [emission] capacity	5, 7
RC	relative charge capacity	5
R_i	ideal gas constant	5
$\mathbf{R_n}$	neutronic radius	4
$\mathbf{R_p}$	relative charge capacity (constant pressure)	5
$\mathbf{R_T}$	gas constant (temperature dependent)	5
R_∞	Rydberg's wave number	4
R_γ	Rydberg's universal constant (energy)	4
t	time	6.10.1 & 2, 6.10
$\mathbf{t_n}$	neutronic period	3
$T_?$	Temperature (key)	4.5.1 to 4
$\mathbf{T_n}$	neutronic temperature	3
v	velocity	4.5.1 to 4
V	electrical voltage	8
V	volume	6.10.1 & 2
X	heat coefficient (velocity)	5
$\mathbf{X_R}$	heat coefficient (orbital radius)	5
Y	temperature coefficient	4
ε_o	permittivity of a vacuum	4
λ	wavelength	6.10.1 & 2, 6.10
μ', μ_o	magnetic constant	4
ρ	density	6.10.1 & 2
Σ	universal constant	3
φ	coupling ratio	4
Ω	electrical resistance	8
ξ_m	mass ratio	4
ξ_v	velocity ratio	4
Table 5.1b		

Symbol	Description
Suffix:	
e	electron
n	neutronic
n	atomic shell number
p	proton
p	constant pressure (heat & charge capacity)
t	constant temperature (heat & charge capacity)
u	ultimate
v	constant volume (heat & charge capacity)
Atomic: temperature, velocity & orbital radius:	
c	cold
m	mean Planck value
n	neutronic
o	minimum Planck value
Modifier:	
N	Newton
P	Planck

Table 5.2c

5.3 Primary Constants

The Tables in this Chapter provide the meaning of the symbols previously listed in Chapter 5.2. In the following Chapters, the constants are grouped according to their properties; general, heat, charge, etc.

There are very few **primary constants**, i.e. those that we must take for granted and on which *all* others are based; these are listed below

Symbol	Value	Table(s)
m_e	9.1093897E-31	kg
The mass of an electron (refer to Chapter 6.7)		
R_n	**2.81793795383896E-15**	m
The neutronic radius (refer to Chapter 6.11.10)		
e	1.60217648753E-19	C
Elementary charge unit (refer to Chapter 6.11.6)		
t_n	**5.90596121302193E-23**	s
Neutronic period		
T_n	**623316124.717178**	K
Neutronic Temperature		
ξ_m	1836.15115053207	
The mass ratio $\{m_p/m_e\}$		
ξ_v	1722.0458764934	
The velocity ratio $\{c/v_o\}$ (refer to Chapter 6.11.5)		
Σ	3E-91 (exact)	m^6
Universal constant (refer to Chapter 6.11.1)		

Table 5.3

Important Note: *The magnetic field constant (B) is not included here as it is simply the reciprocal of the relative Charge Capacity (RC: Table 5). When using 'B' in a formula, such as Lorentz's magnetic force formula, remember: B = 1/RC*

New properties, i.e. those not currently known or used, are highlighted in the following Tables in **bold text**.

5.4 General Physical Constants

Symbol	Formula	Value	Units
G	$a_o.c^2 / \boldsymbol{\rho_u}$	6.67359232004334E-11	$m^3 / s^2.kg$
Newton's gravitational constant (per m³)			
k	$c^2.\mu'$	8.98755184732667E+09	$J.m / C^2$
Coulomb's constant [for an electron] (refer to Chapter 6.11.4)			
k'	k / ξ_m^2	2.6657815048876E+03	$J.m / C^2$
Coulomb's constant for a proton			
φ	$G.m_e.m_p / k.e^2$	4.40742111792334E-40	
Coupling Ratio			
μ'	$\mathbf{R_n}.m_e/e^2$	1E-07	$kg.m / C^2$
Magnetic constant (fundamental)			
μ_o	$4\pi.\mu'$	1.25663706143592E-06	$kg.m / C^2$
Magnetic constant			
ε_o	$1 / \mu_o.c^2$	8.85418775855161E-12	$C^2 / J.m$
Permittivity of a vacuum (e.g. within an atom)			
h	$½.\mathbf{R_n}.m_e.c.\xi_v$	6.62607174469163E-34	$kg.m^2/s$
Planck's constant (resolved into its component parts)			
ℏ	$h / 2\pi$	1.05457207144921E-34	$kg.m^2/s$
Planck's constant (modified by Dirac)			
h'	$½.\mathbf{R_n}.m_e.c^2$	1.15353857232684E-28	$J.m$
Modified Planck's constant			
R_∞	$1 / a_o.\xi_v$	1.09737269561359E+07	$/m$
Rydberg's wave number			
R_y	$\mathbf{R_n}/a_o . ½.m_e.c^2$	2.17987197684936E-18	J
Rydberg's universal constant for the energy of an electron			
α	$e^2 / 4\pi$	2.04272942122269E-39	C^2
Fine structure constant			
X	$\underline{T}_n/c^2$	6.9353271647894E-09	$K.s^2/m^2$
Velocity constant			
X_R	$\underline{T}_n.R_n$	1.75646616508035E-06	$K.m$
Radial constant			
Y	$\sqrt[3]{[½.\xi_v]}$	9.51345439232503	
Temperature coefficient			
e'	$e.\xi_v.\sqrt{[\underline{T}/\underline{T}_n]}$	2.75902141376572E-16	
Proton charge			
Table 5.4a			

PHILOSOPHIÆ NATURALIS PRINCIPIA MATHEMATICA Revision IV

Atomic property constants (refer to Chapter 6.10 for particle properties):

Symbol	Formula	Value	Units
m_e			kg
Mass of an electron (refer to Chapter 5.3)			
m_p	$m_e . \xi_m$	1.672621637830E-27	kg
Mass of a proton			
m_n	$m_e + m_p$	1.6735325768E-27	kg
Mass of a neutron			
a_o	$R_n . (\xi_v / 4\pi)^2$	5.2917721067E-11	m
Rydberg's radius			
R_c	$R_n . \xi_v^3$	1.43901585166681E-05	m
Cold orbital radius (refer to Chapter 3.3.4)			
R_o	$R_n . \xi_v^2$	8.3564315638157E-09	m
Planck minimum orbital radius (refer to Chapter 3.3.4)			
R_m	$R_n . \xi_v$	4.85261843362263E-12	m
Planck mean orbital radius (refer to Chapter 3.3.4)			
R_n			m
Neutronic radius (refer to Chapter 3.3.4)			
v_c	$v_o . \sqrt{[R_o/R_c]}$	4195.20925990715	m/s
Electron cold velocity (refer to Chapter 3.3.4)			
v_o	$c . \sqrt{[R_n/R_o]}$	174090.866621084	m/s
Electron minimum Planck orbital velocity (refer to Chapter 3.3.4)			
v_m	$\sqrt{[c.v_o]}$	7224342.80705004	m/s
Electron mean Planck orbital velocity (refer to Chapter 3.3.4)			
c	$2\pi . R_n / t_n$	299792459	m/s
Electron neutronic velocity (refer to Chapter 3.3.4)			
T_c	$X . v_c^2$	0.122060237421696	K
Cold temperature (refer to Chapter 3.3.4)			
T_o	$X . v_o^2$	210.193328535837	K
Planck minimum temperature (refer to Chapter 3.3.4)			
T_m	$X . v_m^2$	361962.554671561	K
Planck mean temperature (refer to Chapter 3.3.4)			
T_n	$X . c^2$		K
Neutronic temperature (refer to Chapter 3.3.4)			
e	exp(1)	2.71828182845905	
Natural logarithm			
Table 5.4b			

5.5 Universal Heat & Charge Capacities

Symbol	Formula	Value	Units
k_B	$m_e.c^2 / \underline{Y}.\underline{T}_n$	1.38065156E-23	J/K
Boltzmann's constant (refer to Chapter 6.11.13)			
R_i	$k_B.N_A$	8.31447876657891	J / K.mol
Ideal gas constant			
RC	e/m_e	1.75881869180545E+11	C/kg
Relative charge capacity			
Rc	$\sqrt{[\,G/k\,]}$	8.61706029887134E-11	C/kg
Charge [emission] capacity			
R_a	R_i / RAM		J / kg.K
Specific gas constant			
R	$m.R_a$	1.38065156E-23	J/K
Gas constant			
R_p	c_p.RAM	20.7861969164473	J / K.mol
Gas constant; R_i multiplied by 2.5			
R_T	$RAC.q_p.\text{Ln}(\underline{T})$ $R_i.\text{Ln}(N_t)$		J / K.mol
Gas constant ($R_T = R_i$ when $N_t = e$ & $\underline{T}$ = 1.49182469764127 K)			
C_t	$m.c_t$		J/K
Heat capacity (constant *temperature*)			
C_V	$m.c_V$		J/K
Heat capacity (constant *volume*); C_t multiplied by 1.5			
C_p	$m.c_p$		J/K
Heat capacity (constant *pressure*); C_t multiplied by 2.5			
Q_t	$e.\mathbf{q_t}$		J/K
Charge capacity (constant *temperature*); also equal to **R** & C_t			
Q_V	$e.\mathbf{q_V}$		J/K
Charge capacity (constant *volume*); **Q_t** multiplied by 1.5; also equal to C_V			
Q_p	$e.\mathbf{q_p}$		J/K
Charge capacity (constant *pressure*); **Q_t** multiplied by 2.5; also equal to C_p			

Table 5.5a

5.5.1 Microstates

Symbol	Formula	Value	Units
N_t	$\exp(c_p.L_n(T) / R_a)$ $\exp(q_p.L_n(T) / R_a)$ $\exp(2.5 . L_n(T))$		
Microstate (constant *temperature*)			
N_V	c_v / R_a q_v / R_a		
Microstate (constant *volume*); N_t multiplied by 1.5			
N_p	c_p / R_a q_p / R_a		
Microstate (constant *pressure*); N_t multiplied by 2.5			
Table 5.5b			

5.6 Specific Heat Capacities (particles)

Symbol	Formula	Value	Units
RAM$_e$	$m_e . N_A$	5.4858031839070700E-07	kg/mol
Relative atomic mass of an electron			
RAM$_p$	$m_p . N_A$	1.00727638277235E-03	kg/mol
Relative atomic mass of a proton (also the RAM of an hydrogen atom)			
R_{ae}	R_i / RAM$_e$ = k_B / m_e	1.51563563034308E+07	J / kg.K
Specific gas constant for an electron			
R_{ap}	R_i / RAM$_p$ = k_B / m_p	8.25441647276088E+03	J / kg.K
Specific gas constant for a proton			
c_{et}	k_B / m_e	1.51563563034305E+07	J / kg.K
Specific heat capacity for the electron (constant *temperature*)			
c_{eV}	1.5 . c_{et}	2.27345344551458E+07	J / kg.K
Specific heat capacity for the electron (constant *volume*)			
c_{ep}	c_{et} + c_{eV}	3.78908907585763E+07	J / kg.K
Specific heat capacity for the electron (constant *pressure*)			
c_{pt}	k_B / m_p	8.25441647276074E+03	J / kg.K
Specific heat capacity for the proton (constant *temperature*)			
c_{pV}	1.5 . c_{pt}	1.23816247091411E+04	J / kg.K
Specific heat capacity for the proton (constant *volume*)			
c_{pp}	c_{pt} + c_{pV}	2.06360411819018E+04	J / kg.K
Specific heat capacity for the proton (constant *pressure*)			
C_t	$m_e . c_{et}$ $m_p . c_{pt}$	1.38065156E-23	J/K
Heat capacity (constant *temperature*); equal to **R** & **Q$_t$**			
C_V	$m_e . c_{eV}$ $m_p . c_{pV}$	2.07097734E-23	J/K
Heat capacity (constant *volume*); equal to **Q$_v$**			
C_p	$m_e . c_{ep}$ $m_p . c_{pp}$	3.4516289E-23	J/K
Heat capacity (constant *pressure*); equal to **Q$_p$**			

Table 5.6

PHILOSOPHIÆ NATURALIS PRINCIPIA MATHEMATICA Revision IV

5.7 Specific Charge Capacities (particles)

Symbol	Formula	Value	Units
RAC$_e$	$e.N_A$	96485.3317942158	C/mol
Relative atomic charge of an electron (also equal to the Farad)			
RAC$_p$	$e'.N_A$	1.77161652983418E+08	C/mol
Relative atomic charge of a proton (also the RAC of an hydrogen atom)			
R$_{ce}$	R_i / **RAC**$_e$	8.61735002820125E-05	J / C.K
Specific gas constant for an electron			
R$_{cp}$	R_i / **RAC**$_p$	4.69315939796359E-08	J / C.K
Specific gas constant for a proton			
q$_{et}$	k_B / m_e	8.61735002820123E-05	J / C.K
Specific charge capacity for the electron (constant *temperature*)			
q$_{ev}$	1.5 . **q**$_{et}$	1.29260250423019E-04	J / C.K
Specific charge capacity for the electron (constant *volume*)			
q$_{ep}$	**q**$_{et}$ + **q**$_{ev}$	2.1543375070503E-04	J / C.K
Specific charge capacity for the electron (constant *pressure*)			
q$_{pt}$	k_B / **e'**	4.69315939796358E-08	J / C.K
Specific charge capacity for the proton (constant *temperature*)			
q$_{pv}$	1.5 . **q**$_{pt}$	7.0397390969454E-08	J / C.K
Specific charge capacity for the proton (constant *volume*)			
q$_{pp}$	**q**$_{pt}$ + **q**$_{pv}$	1.1732898494909E-07	J / C.K
Specific charge capacity for the proton (constant *pressure*)			
Q_t	e.**q**$_{et}$ e'.**q**$_{pt}$	1.38065156E-23	J/K
Charge capacity (constant *temperature*); equal to **R** & C_t			
Q_V	e.**q**$_{ev}$ e'.**q**$_{pv}$	2.07097734E-23	J/K
Charge capacity (constant *volume*); equal to C_V			
Q_p	e.**q**$_{ep}$ e'.**q**$_{pp}$	3.4516289E-23	J/K
Charge capacity (constant *pressure*); equal to C_p			

Table 5.7

5.8 Electricity

Apart from the Farad, no values are provided for the following electrical properties because the Amp, Volt, Ohm and Henry are now redundant.

Symbol	Formula	Value	Units
A	$e.f$		C/s
Electrical current (Coulomb flow-rate)			
V	PE/e		J/C
Electrical voltage (potential energy per coulomb)			
Ω	$V/A = PE / f.e^2$		$J.s/C^2$
Electrical resistance (momentum over distance per Coulomb squared)			
H	$\mu_o.R$		$kg.m^2/C^2$
Henry; unit of mutual inductance			
F	$e.N_A$	96485.3317942156	C/mol
Farad; unit of electrostatic capacitance (equal to **RAC$_e$**)			
P	$V.A = PE.f$		J/s
Power (Watt)			

Table 5.8

PE is the potential energy between a proton and its orbiting electron

6 Support

A mathematical and descriptive explanation for all the physical constants and scientific discoveries along with the reasons why Relativity and Quantum Theory must now be discarded.

6.1 Proof of the Orbital Model

The four principal agents for the theories of planetary motion were Copernicus, Kepler, Galileo and Newton. Between them, they defined the behaviour of orbiting satellites, moons and planets that remain valid even today.

6.1.1 Nicolaus Copernicus (1473 to 1543)

Copernicus stated that; contrary to religious doctrine, the sun does not orbit the earth, but all the planets in the solar system orbit the sun. He was so concerned for his safety regarding this claim, however, that he arranged for the publication of his findings to be deferred until after his death (1543).

6.1.2 Johannes Kepler (1571 to 1630)

Kepler used Tycho Brahe's (1546 to 1601) observational data to show that the planets not only orbited the sun, just as Copernicus had previously claimed, but that their orbital paths were ellipses. Kepler also stated that the time taken to traverse between any two points (refer to Chapter 2.2.2; Fig 5) on this elliptical curve is proportional to the swept area:

i.e; $t_1/A_1 = t_2/A_2$

Whilst he did not provide a mathematical proof for his swept area theory, he understood it. It was later confirmed by Isaac Newton (below).

6.1.3 Galilei Galileo (1564 to 1642)

Galileo is best known for his physical evidence of celestial bodies (moons) orbiting other planets, revealed in his book; Dialogue Concerning the Two Chief World Systems (frequently referred to as the 'Dialogue'), therein declaring Copernicus correct and finally quashing over a thousand years of religious dogma that stated all celestial bodies orbit the earth. In return for his findings, he was put under permanent house arrest, but only after being threatened with death if he didn't recant this claim.

However, it was during his confinement that Galileo completed his most important work, his laws of motion, one of which states that a body fired from the surface of the earth would follow a parabolic curve back to its surface.

This claim may be demonstrated by comparing Galileo's mathematically correct parabola with a projectile trajectory calculation:

$x(t) = A.t + B$

$y(t) = C.t + D - \tfrac{1}{2}.g.t^2$

If B and D are zero {i.e. v occurs at t = 0}:

$x(t) = v.\text{Cos}(a).t$

$y(t) = v.\text{Sin}(a).t - \tfrac{1}{2}.g.t^2$

Where:
v = initial velocity
A = initial horizontal velocity {i.e.; A = v.Cos(a)}
B = offset horizontal distance from t = 0
C = initial vertical velocity {i.e.; C = v.Sin(a)}
D = offset vertical distance from t = 0

Fig 34 shows the projectile trajectory (curve) superimposed on two alternative parabolic curves, one of which passes through the same latus rectum and the other being the best parallel match.

Whilst the parabolic path is not strictly correct, it is stunningly close, demonstrating that given the limited information and facilities available to Galileo at his time, he was a very capable mathematician.

PHILOSOPHIÆ NATURALIS PRINCIPIA MATHEMATICA Revision IV

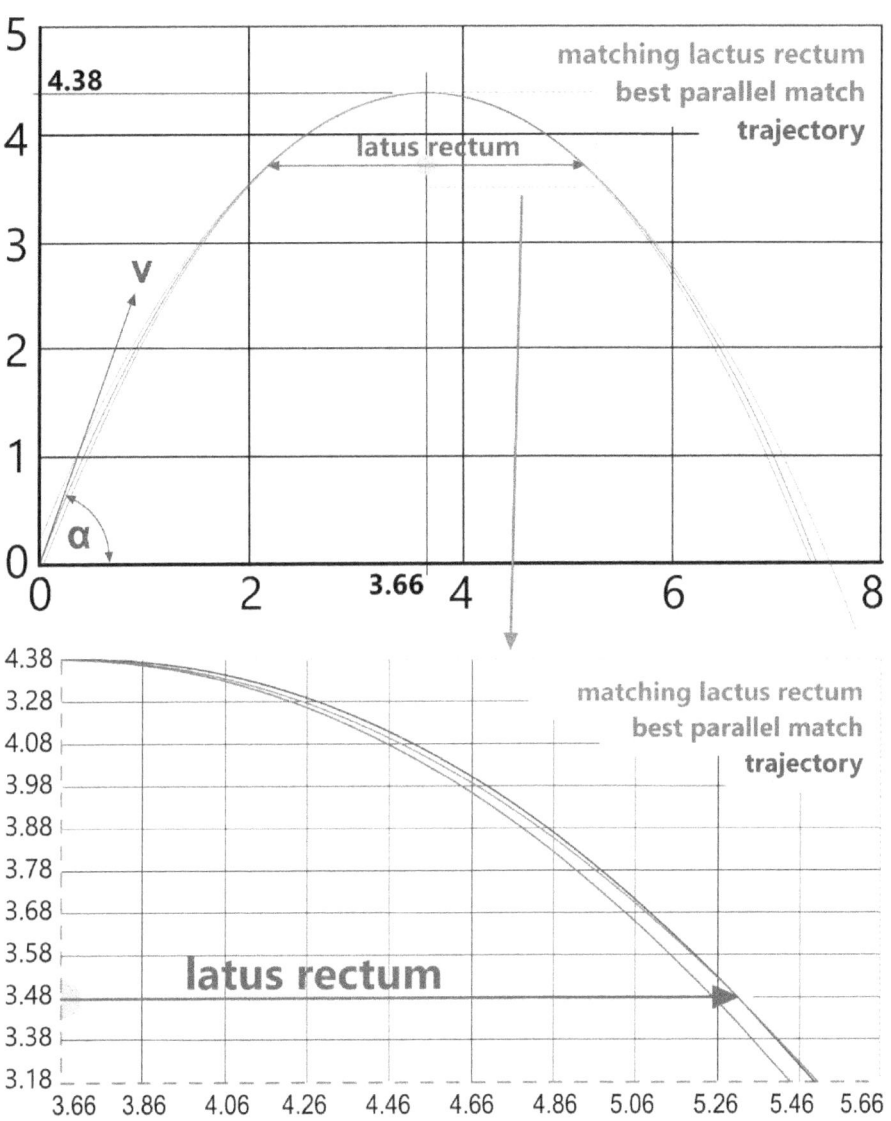

Fig 34

6.1.4 Isaac Newton (1642 to 1727)

Along with [his explanation for] gravity, Newton used [his creation of] calculus to mathematically prove the theories previously generated by Copernicus, Kepler and Galileo. In 1687 he published his results under the heading Philosophiæ Naturalis Principia Mathematica (first of three issues), probably the most important scientific work ever produced.

In his Principia, Newton discusses the alternative curves that describe the elliptical paths followed by an orbiting body. However, the parabolic and hyperbolic curves can only be responsible for paths followed by a body (e.g. a galactic comet) travelling towards a force centre from well outside its influence, sufficiently close to fall under its influence, pass around the force centre and then travel back out of its influence. A complete orbit, i.e. that of a satellite must be an ellipse.

As a result of this work, Newton defined the fundamental relationship (G) between attracting bodies in which the gravitational force (F) is directly proportional to the inverse of the square of the distance (R) between the attracting bodies (refer to Chapter 6.1.4.6).

$$F = \frac{G.m_1.m_2}{r^2}$$

Fig 35

Whilst a value for 'G' was never established by Newton, despite it being of special importance to his theories, it has been estimated many times since the publication of Principia, varying between 6.67E-11 and 6.76E-11 N.m²/kg² (refer to Chapter 6.11.2)

The minimum and maximum radial distances between the earth and sun (Fig 36: @ A & B respectively) are assumed to be as defined in the Earth-Sky fact sheet (https://earthsky.org/). Therefore, using Newton's theories and true value for 'G' (refer to Chapter 6.11.2), the principal properties of the earth's orbit are as follows:

a = 1.495945981E+11m (R + R)/2)
b = 1.495737135E+11 {√[a².(1-e²)]}
e = 0.01670914665 {a.e² + R.e + R - a = 0}
p = 1.495528319E+11m {a.(1-e²)}
f = 1.47095E+11m {a.(1-e)}
x' = 2.499598078E+09m {a-f}
R = 1.47095E+11m to 1.520941962E+11m
F = 3.658178805E+22 to 3.421649078E+22N (refer to Chapter 6.1.4.6)
v = 30286.008788376 to 29290.53557m/s (refer to Chapter 6.1.4.9)

Newton's creation of Calculus allowed him to generate formulas for non-linear versions of Galileo's relationships for distance (s), time (t), velocity (v) and acceleration (a) as follows:
s = ut + ½at²
δs/δt = v = u + at
δ²s/δt² = δv/δt = a

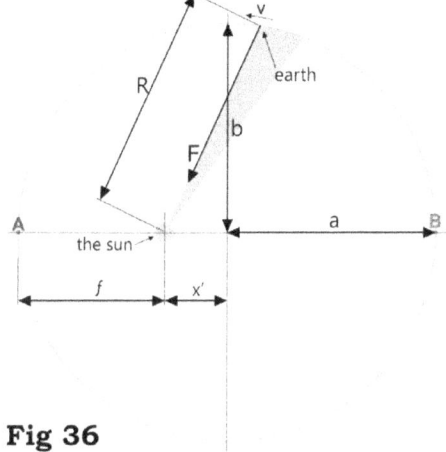

Fig 36

6.1.4.1 Proof (elliptical orbits)

By applying calculus, Newton was able to generate the non-linear formulas necessary to complete his theories concerning the elliptical (conic) path of orbiting bodies, which was proven as follows:

Assume an ellipse and the planet is passing the x-axis @ 'A' (y = 0) (Fig 37)
x component = R {a}
y component = v/ω {b}

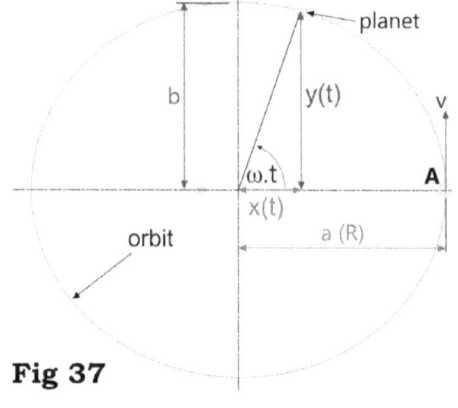

Fig 37

$x(t) = R.\sin(\omega.t)$
$y(t) = (v/\omega).\cos(\omega.t)$

From: $\sin^2(\omega.t) + \cos^2(\omega.t) = 1$

$y(t)^2 / (v/\omega)^2 = 1 - \sin^2(\omega.t)$
$\sin^2(\omega.t) = 1 - y(t)^2 / (v/\omega)^2$

$x(t)^2 / R^2 = \sin^2(\omega.t) = 1 - y(t)^2 / (v/\omega)^2$

$x(t)^2 / R^2 = 1 - y(t)^2 / (v/\omega)^2$
$x(t)^2 / R^2 + y(t)^2 / (v/\omega)^2 = 1$

An ellipse!

6.1.4.2 Euclidean Geometry (equal areas)

Whilst Kepler had already predicted the equal-swept-area-with equal-orbital-time theory, it had still not been mathematically proven by the time Newton was writing his Principia. Newton did this using Euclidian geometry.

The areas of each triangle in Fig 38; A_1, A_2 & A_3 are all equal if the base widths; x_1, x_2 & x_3 are equal, which can be proven as follows:

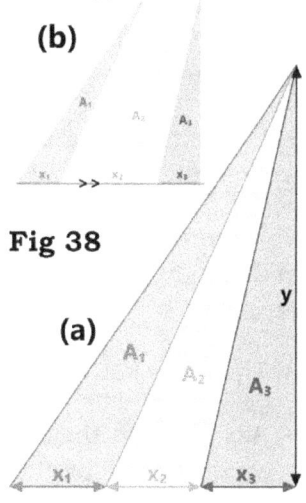

Fig 38

Let $y=6$ and x_1, x_2 & x_3 all equal 3 (a)
The area of a triangle: $A = x.y/2$
$A_1 = 6 \times 3 \div 2 = 9$
$A_2 = 6 \times (3+3) \div 2 - A_1 = 9$
$A_3 = 6 \times (3+3+3) \div 2 - A_1 - A_2 = 9$
Therefore, all the areas are equal (i.e. 9)
The same applies to triangles with equal bases between parallel lines (b)

He then applied this to the conservation of energy (Fig 39)

PHILOSOPHIÆ NATURALIS PRINCIPIA MATHEMATICA Revision IV

6.1.4.3 Proof (conservation of energy & equal time-swept area)

Newton's proposition diagram for his proof of Kepler's 'equal-areas-equal-time' theory is shown in Fig 39, where the following instructions describe its construction (my words):

1) Divide time [of orbit] into equal parts [represented by equal swept areas {triangles}]

2) Assume the line A-B describes the linear path of the body if unconstrained by gravitational attraction

3) The same body would then continue to B-c

4) Assume that the body is attracted by a central-force (S) and diverted from its right line (B-c) in a direction parallel to V-B as far a C

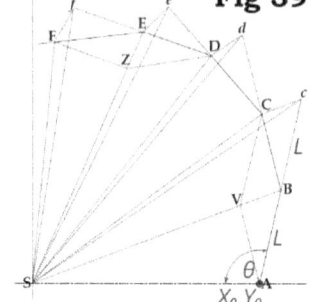

Fig 39

5) Continue to generate similar triangles (S-A-B) following the points D, E, F, etc.

Note: The dimensions L, θ, X_o & Y_o in Fig 39 were not part of Newton's original drawing. They have been added by me in order to assist with the correlation between Figs 39 to 41

Newton was therefore stating that all swept areas (triangles SAB, SBC, SCD, SDE, SEF, etc.) must be equal.

The difficulty in generating the above diagram is knowing how far along the line C-c that C occurs in order to ensure that each subsequent area remains equal.

$$h = r_2 . Tan(\varepsilon_2)$$
$$R_2 = 2.A/h$$
$$\alpha_2 = [h/R_2 . Sin(½\pi - \varepsilon_2)]$$
$$X_2 = R_2 . Cos(\Sigma\alpha)$$
$$Y_2 = R_2 . Sin(\Sigma\alpha)$$

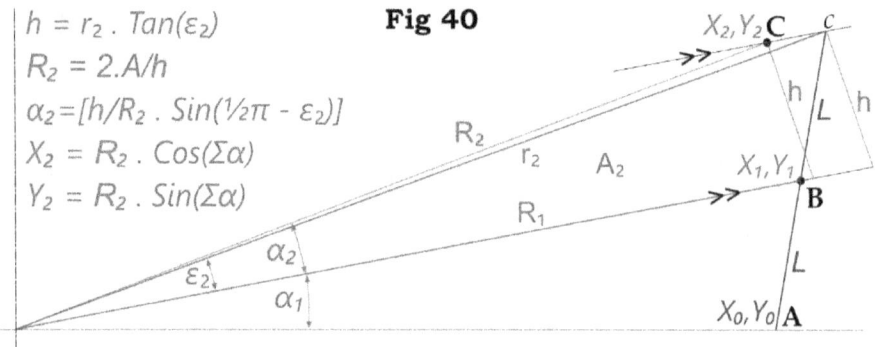

Fig 40

254

This can be achieved using the process described in Fig 40, where the blue variables are entered (X_0, Y_0, L, θ), all the green variables (X_1, Y_1, R_1, r_2, A_2, $ε_2$, X_0, a_1) can be easily calculated using the blue variables and the red variables (h, R_2, a_2, X_2, Y_2) may be determined using the formulas provided.

Newton claimed that if you reduce length L 'in infinitum' and join up the dots (X,Y co-ordinates) you produce a curved line thereby demonstrating that a centripetal force is continually acting on the body in the direction of the force centre and the triangular areas will always be proportional to the time passed by the body traversing each triangle; QED

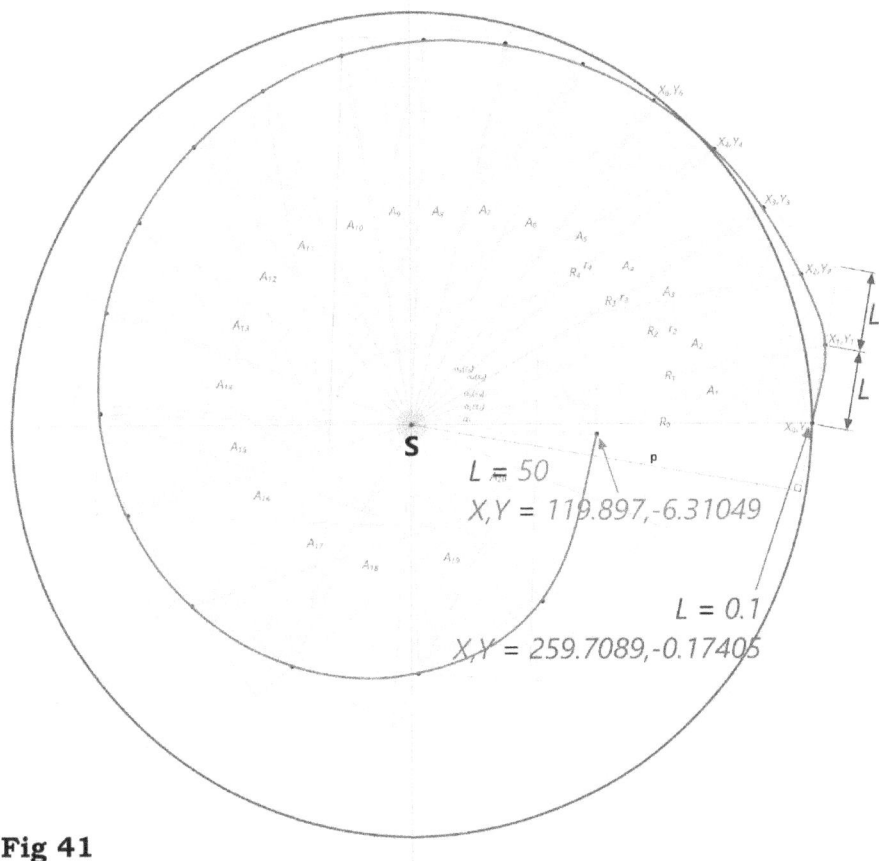

Fig 41

This argument is less easily seen from his words and his fairly simple diagram (Fig 39) than if you actually complete his diagram and repeat it for ever smaller values of L through to 360° (see below)

PHILOSOPHIÆ NATURALIS PRINCIPIA MATHEMATICA Revision IV

A calculation was carried out using the following input data:
X_0, Y_0 = **260,0**
L = 50
θ = 100°

As can be seen in Fig 41, the diagram does indeed produce a curve, exactly as Newton claimed

... and following this through a sequence of diminishing values for L from 50 to 0.1, the following X,(Y) co-ordinates are achieved immediately prior to reaching 360°:

L	X	(Y)
50	119.8970215	(-6.310487628)
25	179.800507	(-11.66860541)
10	229.5181732	(-9.486414926)
1	257.0676077	(-0.349331185)
0.1	**259.7089445**	**(-0.174048539)**

Note: the 'Y' co-ordinate is in parenthesis because it is simply a resultant. The trend is demonstrated by the 'X' co-ordinate.

... from which it isn't difficult to anticipate where X (& Y) will end up if L is diminished in infinitum [i.e.: X(,Y) = **260(,0)**], making the final shape a circle and thereby proving that:

a) the path of the body is continuous (conservation of energy and angular momentum)

b) the orbital time passed by the body is proportional to the swept area (triangle)

c) Newton's calculus can be used to determine the properties of the path {'in infinitum'}

This result does not mean that the orbital path is circular, simply that it is continuous.

The orbital path is calculated using the procedure provided in Chapter 3.2.2 above.

Corollary 1

Newton's first Corollary (to the above proof) states that the velocity of the body (v), represented by L, at positions A, B, C, D, E, F, etc. (Fig 39) is inversely proportional to the perpendicular distance of its tangent from the force centre (Fig 41; p {just to the right of "L = 50"})

Newton also stated that; v multiplied by p is a constant, i.e. his constant of motion (h), which is the angular momentum without the mass component.

Using the above 'in infinitum' argument it can be seen in the following table where these calculations have been carried out for successively reduced values of L between the start and end of the orbit (h_n @ 0°, h < 360°), 'h' does indeed become a constant:

L	h_n	h
50	2201.214	7485.049
25	580.0837	1196.551
10	95.12179	123.8112
1	0.963516	0.992077
0.1	**0.0970**	**0.0974**

Newton's constant of motion 'h' is not to be confused with the perpendicular distance 'h' shown in Fig 40; they are neither the same nor in any way connected.

6.1.4.4 Centripetal Force

Centrifugal acceleration (according to Christiaan Huygens {1629 to 1695}):
$a = R.\omega^2$
where $\omega = 2.\pi/t$
$a = \sqrt{[(R.\omega^2)^2 + (R.\alpha)^2]}$
with constant angular momentum; $\alpha = 0$
$a = R.\omega^2$
$a = R.(2.\pi/t)^2$

Centrifugal force:
$F = m.a$
$F = m.R.(2.\pi/t)^2 = 4.\pi^2.m\ (R/t^2)$

Through his inverse rules, Newton shows that the centripetal force (F) between the orbiting body and the force-centre (Fig 42);
$F = SP^2 . QT^2 / QR$

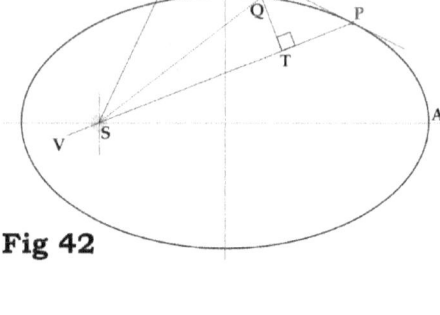

Fig 42

$PR = v^P . \delta t$
where; v^P is the velocity of the body at P and δt is the time taken for the body to travel from P to Q

$QR = (F^P / 2m).\delta t^2$
where; F^P is the centripetal force on the body at P
$F = QR . (2m / \delta t^2)$
where; F^P is the centripetal force on the body at P

$\delta t = PR/v^P = PR / (h/SY) = PR . SY / h$
where h is Newton's constant of motion (see Corollary 1 above)

Therefore, the centripetal force (F) can be calculated as follows:
$F = QR . (2.m / (PR . SY / h)^2)$
$\quad = QR . (2.m / (PR^2 . SY^2 / h^2))$
$\quad = QR . (2.m.h^2 / (PR^2 . SY^2))$
$\quad = QR.2.m.h^2 / PR^2.SY^2$

Newton preferred the calculation in geometric form by setting $2.m.h^2$ as a constant (k):
$F = k.QR / (PR.SY)^2$

6.1.4.5 Distance Between A Satellite & Its Force-Centre (R)

The separation (distance) between an orbiting body and its force centre, can be found by using general elliptical equation:
$R = a.(1-e^2) / (1-e.Cos(\theta))$
where 'R' & 'θ' are as shown in (Fig 44) and 'e' is eccentricity

The force centre is not at the centre of an ellipse but at its focus (Fig 42; S)

6.1.4.6 The Inverse Square Law

Proposition XI: *"If a body revolves in an ellipse; it is required to find the law of the centripetal force tending to the focus of the ellipse"*

Using similar geometric arguments as above (Figs 38 to 42) Newton worked out that the force between an orbiting body and its force centre is proportional to the inverse of their separation (the distance between them): $F \propto 1/R^2$ i.e. $F = K / R^2$ (refer to Chapter 2.2; Page 45)
where:
the constant of proportionality: $K = G.m_1.m_2$ i.e. $F = G.m_1.m_2 / R^2$
where G is a constant and m_1 and m_2 are the masses of the force centre and the orbiting body

This same relationship ($F \propto 1/R^2$) also applies to parabolas and hyperbolas as well as the ellipse

The above constant of proportionality (K) can also been written as;
$K = m.h^2/p$
Where 'h' and 'p' are defined in Corollary 1 above and m is the mass of the orbiting body
i.e. $F = (m.h^2/p).(1/R^2)$
In the first formula, you can resolve the problem knowing the mass of the bodies
In the second formula, you can resolve it knowing the velocity and mass of the orbiting body and the parameter of its curve (p)

Both of the above F calculations produce the same result;

e.g. the following centripetal force occurs in the earth's orbit, 0.000175° from the major semi-axis:

G = 6.67359232004332E-11 (gravitational constant)
m_1 = 1.9885E+30 (sun mass)
m_2 = 5.964519768E+24 (earth mass)
R = 1.5209420E+11 (distance between mass centres)
$F = G.m_1.m_2 / R^2$ = **3.421649078E+22** (centripetal)

h = 4.454920463E+15 (constant of motion - see Corollary 1 above)
m = 5.964519768E+24 (earth mass)
p = 1.495528319E+11 (ellipse parameter)
R = 1.5209420E+11 (distance between mass centres)
$F = m.h^2 / p.R^2$ = **3.421649078E+22**

6.1.4.7 Orbital Period

Proposition XV: *"The same things being supposed, I say, that the periodic times in ellipses are as the $3/2^{th}$ power of their greater axes"*

This means that if the major semi-axis of an ellipse is 'a' (Fig 37) and the time taken for a body to orbit the elliptical path is 't' then the relationship between the two is:

$t \propto (2.a)^{1.5}$ or $t^2 \propto (2.a)^3$

Therefore; $t = K \cdot a^{1.5}$

Where K is the constant of proportionality, which is dependent on the properties of the force-centre.

This is actually Kepler's third law

6.1.4.8 Constant of Proportionality

To determine 'K' (the constant of proportionality for $t = K \cdot a^{1.5}$
{refer to Chapter 6.1.4.7}) ...
$K = t^2 / a^3$ {s^2/m^3}
now we know ...
... that the earth travels around the sun in 31558149s
... the earth's semi-major orbital x-axis is 1.495945981E+11m

Therefore:
t^2 / a^3 = **2.974914364E-19** {s^2/m^3}

G = 6.67359232E-11 {$N.m^2/kg^2$ = $kg.m.m^2 / s^2.kg^2$ = $m^3 / s^2.kg$}
m_1 = 1.9885E+30 kg (the mass of our sun)
$1 / m_1.G$ = 7.535546116E-21 {$s^2.kg / m^3/kg$ = s^2/m^3}
2.975944645E-19 ÷ 7.538155846E-21 = 39.47841760436
$\sqrt{39.47841760436}$ = 6.2831853071796 = $2.\pi$

Therefore:
$K = (2\pi)^2 / G.m_1 = (2\pi)^2$ ÷ 6.67359232E-11 ÷ 1.9885E+30
   = **2.974914364E-19** s^2/m^3
i.e.;
$K = (2\pi)^2 / G.m^{fc}$
where m^{fc} is the mass of the force-centre

The above calculation, based upon NASA's data for the sun and the earth's orbit, gives an error margin of 0

6.1.4.9 Alternative Velocity Calculation

A much simpler orbital velocity calculation method is based upon Kepler's 'swept-area = time' rule (Fig 43)

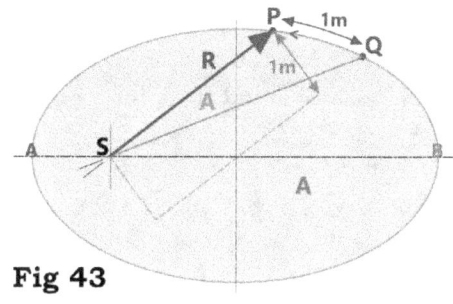

Fig 43

Using the earth's orbit as an example:

Earth's total orbital area (A) is 7.029445371E+22m² and it takes 31558149s (t) to complete

The swept area (A) is equal to ½.R x 1m
The velocity of the orbiting body at any given distance between the centres of mass (P & S) is calculated as follows:
v = 2.A / t.R {m² / s.m = m/s}

By way of verification:

The earth's maximum velocity occurs when R = 1.47095E+11 m (@ A)
v = (2 x 7.029445371E+22) ÷ (31558118.4 x 1.47095E+11)
 = **30286.008788376** m/s (refer to Chapter 4.2.2)
30286.008788376 m/s (calculated using; h=v.R)

The earth's minimum velocity occurs when R = 1.520941962E+11 m (@ B)
v = (2 x 7.02944537126484E+22) ÷ (31558149 x 1.520941962E+11)
 = **29290.5355716777** m/s (refer to Chapter 4.2.2)
29290.5355716777 m/s (calculated using; h=v.R)

The above confirms Kepler's 'swept-area = time' rule and shows that
v ∝ 1/R

or v = k/R
where k = 2.A / t

6.1.4.10 Centrifugal force in an orbiting body

In any orbiting system, the centripetal force, i.e. Newton's gravitational force (Fig 35), must be equal to the orbiting body's centrifugal force, which may be calculated thus (Fig 44):

$F = m_1 . v_1^2 / R$
$Fc = m_1 . v_2^2 / R$
where
$v_2 = \sqrt{[G.m_1 / R]}$

@ the perihelion (perigee) of an ellipse; $Fc = F . f/p = F / (1+e)$
@ the aphelion (apogee) of an ellipse; $Fc = F . p/f = F . (1+e)$

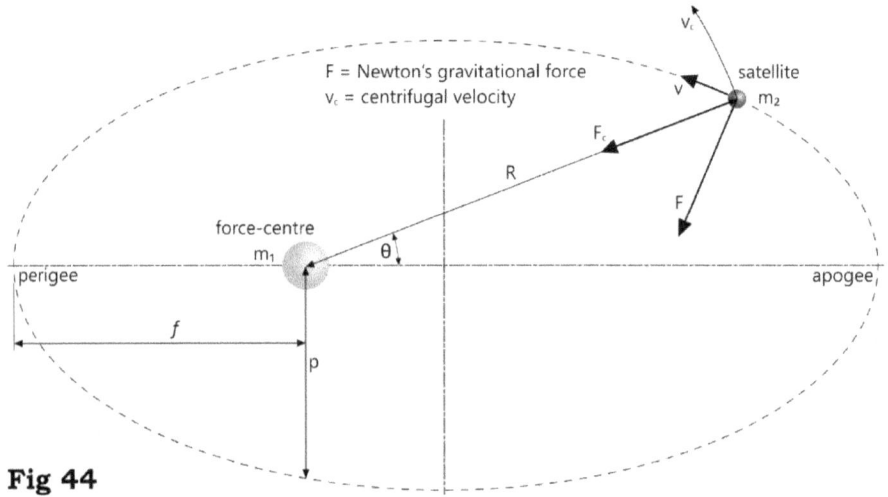

Fig 44

Orbital velocity anywhere in an orbit may be calculated thus:
$v_2 = 2\pi.a.b / R.t$
where: t is the satellite's orbital period

Centrifugal velocity anywhere in an orbit may be calculated thus:
$a = \sqrt{[4/3.\pi]}$
$\zeta = \sqrt{[(f.\text{Sin}(\theta/2)^a + p.\text{Cos}(\theta/2)^a) / (f.\cos(\theta/2)^a + p.\text{Sin}(\theta/2)^a)]}$
$V_c = \zeta.v_2$

PHILOSOPHIÆ NATURALIS PRINCIPIA MATHEMATICA Revision IV

6.1.4.11 Fundamental Laws of Orbital Motion

1) Every orbital system must have a force-centre and at least one satellite

2) A force-centre's mass defines its satellite's orbital shapes and periods

3) Satellite orbits define a force-centre's spin

4) Sub-satellite orbits and force-centre spin define a satellite's spin

5) Sub-satellites have no effect on the force-centre

6) Satellites may be swapped between orbits without altering orbital shapes and periods; e.g. Jupiter may replace Earth and Jupiter would follow the same orbital path that Earth previously followed and would orbit in 365¼ days

6.2 The Problem with Relativity

For reasons of brevity, I shall refer to the theories of general and special relativity as 'Relativity' and the author of these theories as the 'Author'.

These Chapters (6.2.1 to 6.2.6) include numerous reasons why Relativity can no longer be considered appropriate for orbital motion, the most significant of which are:

1) *All* orbits work perfectly without it (i.e. it is unnecessary)

2) It causes many aspects of Newton's laws of orbital motion to fail

3) It was developed simply to support item 4) below

4) The Author misunderstood the meaning of both $E=mc^2$ & light

Just one, or even two, of the following problems (6.2.1 to .6) could be chalked up to coincidence, but all six!

A major concern regarding Relativity is the lack of attention paid to matching units in its formulae. For example, it appears to include the formula: '$R_{ab} - \frac{1}{2}R.g_{ab} = T_{ab}$'; in which length is added to velocity squared which results in time. Even based upon Reimann mathematics, this doesn't make much sense.

It is important to remember:

Whilst it is possible to create a sub-theory to explain any distortion of reality you wish, why would you if there is no need?

When everything in the universe can be explained without the sub-theory, the sub-theory becomes redundant.

Relativity was driven by a desire to explain events that were either unknown or misunderstood. Now that we fully understand the theory behind *all* orbital systems and *light*, Relativity has become redundant, especially as it actually invalidates Newton's laws of orbital motion, that otherwise work perfectly, *in every respect*; irrespective of energy, speed and mass.

It seems clear to me, that Relativity must be declared *'dead in the water'*

6.2.1 Light Deflection

Light is apparently observed to deflect by an angle of 1.75 arc-seconds when passing at or close to the surface of our sun (Fig 45; α).

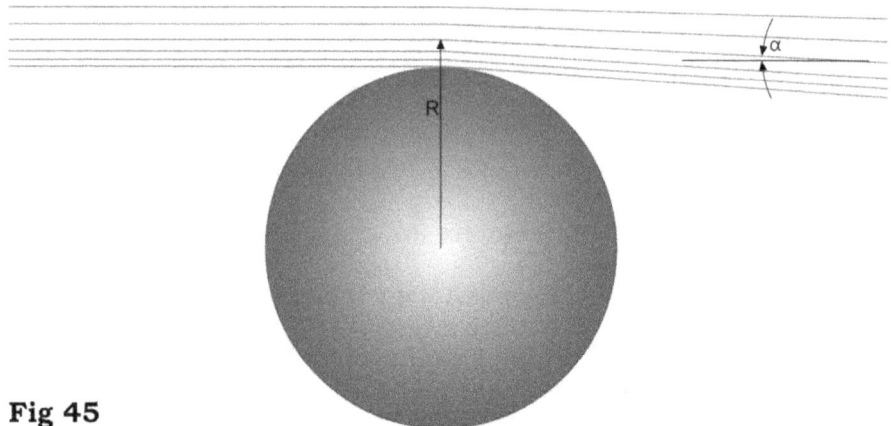

Fig 45

According to the Author, however, Isaac Newton's laws of gravity predict a deflection angle of half this value (0.875)

Relativity is a mathematical description of space-time/gravity distortion around celestial bodies that was developed to support the Author's formula for *observed* light deflection angles, which was based upon 'light-emitting photons' and their susceptibility to gravity.

The problem with the Author's approach is that the light we see all around us is electro-magnetic energy, and therefore possesses no mass, so gravitational laws don't apply (refer to Chapter 6.2.2).

On the other hand, Isaac Newton's gravitational constant (G), which is based upon the properties of Quanta, may be used to define the same deflection angle (α) as follows:

α = Atan(4.a$_o$.V$_s$ / R) {m³.m / m³.m}   #

Where:
$G = a_o.c^2/\rho_u$ {m³ / kg.s² per m³}
$V_s = m_s/\rho_u$ {m³}
m_s = the mass of our sun {kg}
ρ_u = ultimate density {kg/m³}
a_o = Bohr's radius {m}
c = the speed of light in a vacuum {m/s}

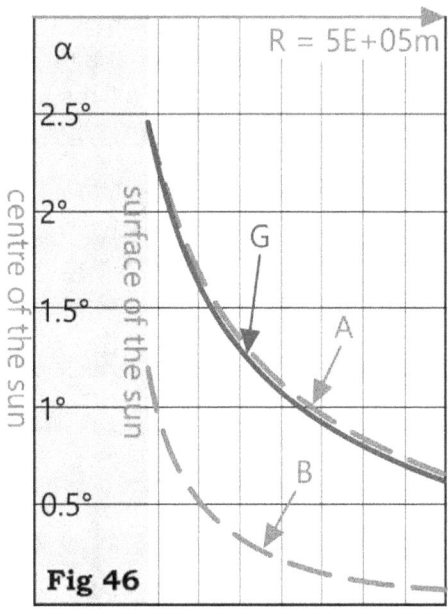

Fig 46

When calculating light deflection angles, the Author reduced the size of our sun by a factor of 5000, retaining its correct mass and increasing its density accordingly. Presumably, this was to raise the calculated angles to a practical value.

Fig 46: Curve **A** shows the variation in 'α' according to **Relativity**, from the surface of the above modified sun to a distance 5E+05m from its centre, which complies with the *observed* deflection value.

Fig 46: Curve **B** shows the equivalent variation in 'α', according to the Author, when using **Newton's laws of gravity** on a *'photon'*

Fig 46: Curve **G**; we can also reproduce the *observed* light deflection angles if we plot the light deflection angles using formula # (above), which is based upon **Newton's gravitational constant**. The difference being that this formula, which is not based upon the gravitational susceptibility of *'photons'*, does not need a sub-theory to justify it.

It is important to understand that it isn't necessary to prove the validity of an alternative theory in order to discredit the original. You only need to demonstrate that the same result may be achieved by applying valid input data to an identical model but without the need for a sub-theory for justification. The above formula does exactly that.

That light travels in waves and not as particles is not new; Christiaan Huygens declared this to be the case in the late 17th century and it was later restated by Faraday, Maxwell and Pauling.

Newton on the other hand declared light to be particles. The Author automatically declared Newton to be correct because he quite rightly held Newton in such high esteem, and therefore used Newton's laws of gravity to deflect [particle] light as it passed celestial bodies. The problem was that this produced incorrect values (Fig 46: Curve **B**), so the Author created a sub-theory (Relativity) to justify his approach.

General relativity, the deformation of space-time, is based upon the inability to use Newton's laws of gravity to predict the deflection of light, which is only a problem if light possesses mass, which it doesn't. This theory is therefore based upon a misunderstanding of the nature of light.

General relativity was also devised because of the Author's disbelief in force-fields, which he called the 'ether'. But whilst there is no such thing as *the ether* (as he understood it) we do know that force fields exist, as anybody holding two magnets close together, but not touching, will know.

6.2.2 The *Speed of Light*

All physicists currently claim that the light we see is emitted by photons.

They also claim that electrons are 'weird beasts' that possess mass and travel in waves, which is the reason we cannot pin them down (uncertainty principle).

This is difficult to understand given that; if the entire electro-magnetic spectrum ranges between <2E-14m and >7m, how can all photons travel at the same velocity. Surely, they must travel at all speeds between >0 to 'c' in order to represent the full electro-magnetic spectrum.

For example; an electron travelling at 1E+06m/s will possess a different energy to one travelling at 1E+08m/s. The electro-magnetic energy (wavelength; e.g. colour) each radiates must, therefore, also be different. And if so, according to Newton's laws of gravitational attraction, different wavelengths of light must be deflected at different angles (refer to Chapter 6.2.1: α).
Which is contrary to the fundamental principle of both theories of Relativity: That "light possesses mass" and all wavelengths of light passing the same celestial body at the same radial distance is deflected at the same angle (α).

Therefore, the light we see cannot be photons, it must be electro-magnetic energy, which possesses no mass and is deflected by magnetic charge.

Not only is it unnecessary to deform space-time around celestial bodies in order to explain light deflection, it is mathematically incorrect to do so.

Special relativity was devised because of the inability to correlate the additive nature of mass-velocity with the non-additive nature of light. This is only a dilemma if light possesses mass, which it doesn't. The *photon* is therefore, also based upon a misunderstanding of the nature of light.

There is no such thing as a photon (refer to Chapter 6.4).

6.2.3 Neutronic Radius (Rn)

The neutronic radius (refer to Chapter 3.5.1.3), which is achieved by an orbiting electron when travelling at 'c', can *only* be explained using Newton's laws of orbital motion and Coulomb's law of electrical force. It occurs in far too many constants (magnetic, permittivity, Rydberg's, Planck's, Coulomb's, Henry's, etc.) to be rejected as a *fundamental physical constant*.

The neutronic radius is also the basis of $E=mc^2$ (refer to Chapter 6.2.5)

The conversion of mass to energy with velocity together with the space-time/gravitational distortion around force-centres as defined in Relativity, would render such an orbital radius impossible. I.e. the electron would be orbiting inside the proton at 'c' and R_n would be incorrect, making magnetic constant, permittivity, Rydberg, Planck, Coulomb, Henry, etc. incorrect, which we know is not the case.

6.2.4 Elliptical Orbits

Relativity is based upon the predication that light possesses mass (refer to Chapter 1.1.1) and that gravity is responsible for its deflection, and because the Author claimed that Newton's laws of gravitational attraction cannot apply to light (refer to Chapter 6.2.1), it was necessary to deform space-time around celestial bodies and artificially modify satellite velocity to account for this problem. But if we apply the relativistic velocity modification to a satellite, such as the earth (Fig 47); $v = v / \sqrt{[1+(v/c)^2]}$ we find that; whilst Newton's laws <u>always</u> work (centrifugal and gravitational acceleration always balance) irrespective of satellite velocity, Relativity fails above <1% of *light-speed*. In fact, according to Relativity an electron can never actually achieve *light-speed*: $c \neq c/\sqrt{2}$

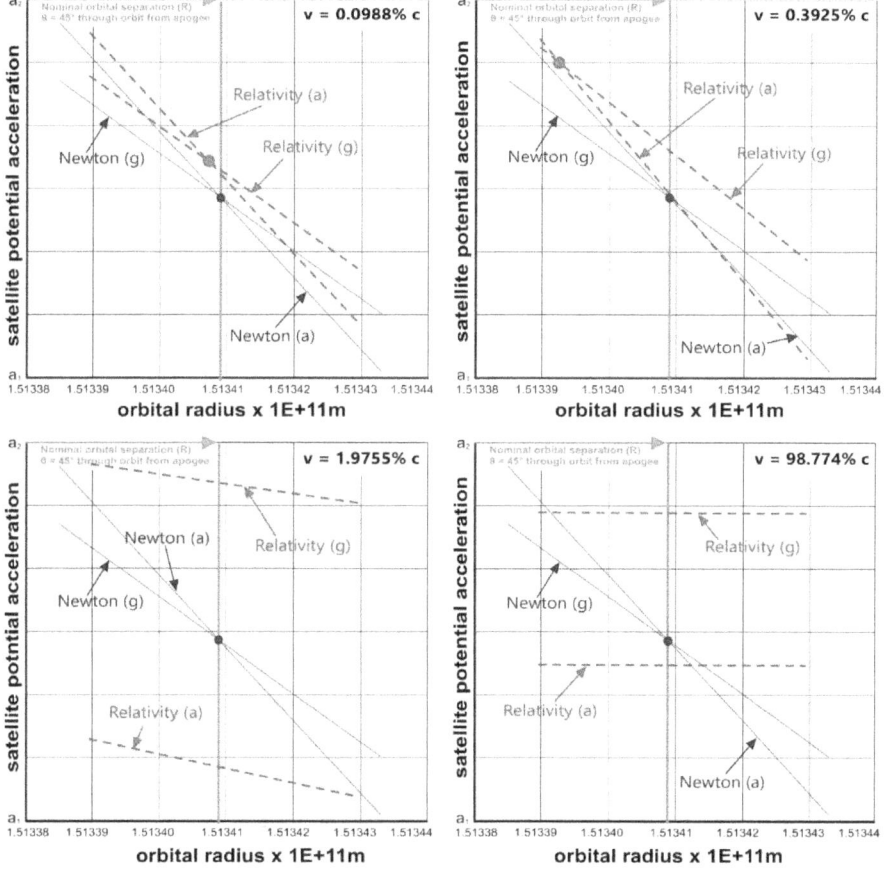

Fig 47

Fig 48 shows the same calculation performed for an electron passing the sun at the *speed of light* at an orbital radius of 765061000m (refer to Chapter 6.2.1; Fig 45) in which Newton's laws of orbital motion function correctly with no modification to elliptical orbits or satellite velocity. Relativism, however, shows gravitational acceleration at *light-speed* is always greater (more than twice) centrifugal acceleration, meaning that a *photon* cannot pass the sun without being absorbed by it; if light possess mass and gravity is responsible for its deflection (as the Author claimed).

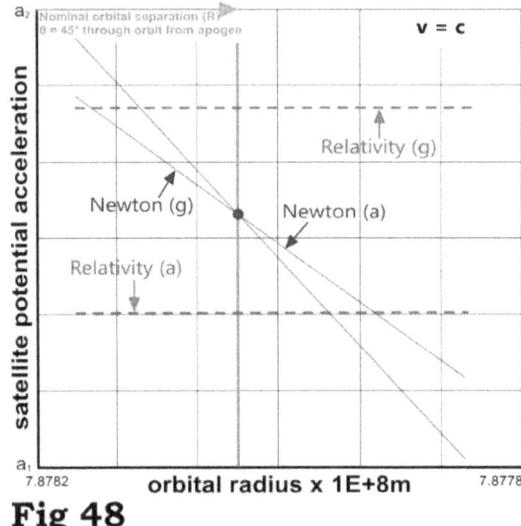

Fig 48

In fact, simply because the Author assumed that light possesses mass, he found it necessary not only to modify space and time but also artificially modify velocity, and even then, his theory fails in that it predicts an electron cannot achieve *light-speed*, destroying his concept of a *photon* (Figs 47 & 48).

Elliptical orbits are an indisputable fact of nature. This has been repeatedly demonstrated since Kepler's discovery in the 17th century. Its mathematical laws show that an exact ellipse is *fundamental* to the constant of [orbital] motion and thereby essential to maintain satellite paths in non-circular orbits. Relativity requires a distortion of this ellipse, rendering the orbital laws unworkable; yet we know that Newton's universal orbits work perfectly irrespective of size, shape and speed (refer to Chapter 4).

Moreover, if we alter time according to Relativity ($T_{AB} = R_{AB} - \frac{1}{2}.g_{AB}.R^\#$) the balance will shift ensuring that external interaction (from other bodies) cannot be rectified naturally; yet they do, as demonstrated in Chapter 2.2.8 and Figs 47 & 48 (Newton)

\# *The units within this formula, which do not match, cannot be reconciled without a sub-theory. Such an anomaly does not occur in any of Isaac Newton's theories.*

It must therefore be concluded that relativity is not only unnecessary; this aspect of the theory is also incorrect.

6.2.5 E=mc²

E=mc² was first postulated by Henri Poincaré towards the end of the 19th century (c = $\sqrt{[E/m]}$), however, he did not explain its physical relevance other than it represented a terminal velocity.

Numerous formulas were contrived in an attempt to rationalise this belief;
m = m$_o$ / $\sqrt{[1\pm(v/c)^2]}$ &   v = 1 / $\sqrt{[1\pm(v/c)^2]}$
but it appears to show that mass is infinite when; v = c or v can never exceed 'c/√2'

During the creation of this publication, I discovered that E=m.v² represents the *potential* energy between a satellite and its force-centre in circular orbits (e.g. atoms), where; PE = -2.KE (refer to Chapter 2.2.3) KE = ½.m.v²; PE = m.v²
and at the speed of light, an electron orbits its proton at radius R_n, which is a fundamental constant (refer to Chapter 6.2.3) where PE = m.c² at the creation of a neutron.

Whilst Poincaré's formula (E=m.c²) does indeed represent a terminal velocity, it refers to the ultimate potential energy between a proton and its electron that is *orbiting* at 'c'. And occurs when the magnetic attractive field energy exceeds centrifugal repulsion energy and the electron combines with its proton to create a neutron. It does not refer to an electron in free-flight travelling at the speed of light, or in fact, kinetic energy of any kind.

That said, electro-magnetic energy can only be radiated whilst an orbiting electron is travelling at less than 'c', which means that no electron can *naturally* achieve this kinetic energy (which it gets from electro-magnetic radiation): E = ½m.c². However, this limitation *does not* mean that an electron, or anything else, cannot travel faster than 'c' if given sufficient energy artificially.

This means that in Relativity, E=mc² has been inappropriately applied to kinetic energy to describe mass-energy variation with velocity, which does not actually occur; mass does not vary with velocity and there is no such thing as mass.

PHILOSOPHIÆ NATURALIS PRINCIPIA MATHEMATICA Revision IV

Combining the theories from Newton, Planck and Poincaré:
Assuming 'm' is a unit of mass of ultimate density: $m = \rho_u$
Newton: $G = a_o.c^2 / m$
Planck: $F = c^4/G$
$F = m.c^4 / a_o.c^2$
$F = E/R$
$E = F.a_o$
$E = m.c^2$ (Poincaré)

The following is an hypothetical argument because black-holes are not sized to trap electrons (refer to Chapter 2.7.2), but the mathematical argument is valid because it describes the relationship between $E=mc^2$ and Schwarzschild's radius.

Kristian Huygens gave us the relationship between acceleration and velocity;
$v^2 = 2.a.R$ (for the PE component)
and Henri Poincaré showed us that $E = m.c^2$, which today is generally believed to represent kinetic relativism but we now know that potential energy is twice kinetic energy in circular orbits (e.g. atoms):
$PE = 2.KE = m.v^2$
If we assume a limiting gravitational energy that will trap an electron travelling at the speed of light, it is probably equivalent to that defined by Henri Poincaré, i.e. for any specified mass; $m.c^2 = m.2.g.R$
where 'g' is the gravitational acceleration at its outer surface (at radius 'R')

Therefore, for any specific gravitational energy, according to: $E = m.c^2$ we should be able to find the associated limiting mass with respect to its ability to emit light:
$c^2 = 2.g/R \rightarrow g = c^2 / 2.R$
$E = m.g.R \rightarrow m.R.c^2 / 2.R \rightarrow \frac{1}{2}.m.c^2$ (kinetic energy at light speed)
i.e. if $2.g.R \geq c^2$ for a given force centre, electrons will have insufficient energy to escape its surface.
if $g = G.m/R^2$ then $2.G.m/R = c^2$ represents the limiting mass

c^2 in this famous equation therefore represents a limiting gravitational acceleration that may be used to define the potential energy required to trap an electron travelling at the speed of light, and the formula becomes:
$E = m.g.R$
where $g.R = c^2$ and the term 'm.g' refers to the potential force on electrons

'E' in this formula is not kinetic energy, it is potential, i.e. Henri Poincaré's famous formula wasn't showing us what has euphemistically become relativism;
between them, Isaac Newton and Henri Poincaré were showing us how to size a [fictitious] black hole, because Newton incorrectly believed that light possessed mass!

$E = m.c^2 = m.2.R.g$
$c^2 = 2.R.g$
$g = G.m/R^2$
$c^2 = 2.R.G.m/R^2 = 2.G.m/R$
$R = 2.G.m/c^2$

Schwarzschild radius of a proton (Rs)
If 'm_p' is the mass of a proton:
$Rs = 2.G.m_p/c^2 = 2 \times 6.67359232E\text{-}11 \times 1.67262164E\text{-}27 \div 299792459^2$
$Rs = 2.48396784934951E\text{-}54$ m

When an electron is orbiting at the speed of light (c), the electron-proton separation radius '$R_n = 2.817729E\text{-}15$ m'
Moreover; $Rs = 2.\varphi.R_n$

This is the point at which an electron comes together with a proton to produce a neutron. The temperature of the electron orbiting at the speed of light is: $\underline{T} = X.c^2 = 620963351.43177$ K

All of which demonstrates that '$E=mc^2$' applies to potential, not kinetic energy. Therefore, electrons (and all matter) in free-flight are not limited to the speed of electro-magnetic energy (e.g. light).

$E=mc^2$ has nothing to do with kinetic energy and mass does not change into energy with increasing speed.
$E=mc^2$ (which was discovered/prophesied by Henri Poincaré) refers to circular orbits where potential energy (PE) is twice kinetic energy ($PE = 2.½.m.v^2$).
At the velocity of light (v = c), a proton and its electron combine to create a neutron (refer to Chapter 6.2.3).

'$E=mc^2$' refers to potential energy, not kinetic energy.

6.2.6 Hades

At the time Relativity was theorised, neither its Author or anyone else was aware of the exigency of force-centres in *every* orbital system or of [planetary] spin theory. The Author therefore misunderstood the effect of galactic population on orbital shapes (refer to Chapter 2.2); hence the misguided invention of dark matter (refer to Chapter 2.7.4).

Moreover, if the Author had known of Hades and the laws of station-keeping (refer to Chapter 2.2.8), he would have realised that the deformation of space-time could not work.

6.3 The Problem with Quantum Theory

For reasons of brevity, I shall refer to Quantum Theory as 'QT' and its author as the 'Author'.

There are numerous reasons why QT can no longer be considered appropriate for the description of atoms, the most significant of which are listed below.

1) Whilst QT cannot explain or describe the behaviour of atoms in terms of what we see, feel and hear in the universe, an atom according to Isaac Newton and Coulomb can do this.

2) QT requires a still-undiscovered unification theory to ensure compliance with Newton's laws of motion.

3) The non-orbital nature of QT electrons means the QT atom cannot generate or emit electro-magnetic energy.

4) QT required the invention of 'string-theory' along with numerous sub-atomic particles (e.g. quarks, fermions, bosons, gluons, etc.) in order to make the atom work, whereas Newton and Coulomb can make the entire universe work with just two particles; the electron and the proton.

5) QT relies on statistics for justification; statistics apply only to the consequences of the laws of nature, never the laws themselves.

6) As is demonstrated in Chapter 4.5, Newton's and Coulomb's theories can be applied to the atomic structure described in this book and therewith mathematically predict its properties.

7) QT's Author needed intimidation to force acceptance of his theories by the scientific community;
"if you aren't profoundly shocked by quantum physics, then you haven't understood it"

8) QT remains unproven after 100 years.

9) It was necessary to invent sub-theories (including the uncertainty principle) to explain why electron location cannot be predicted in QT. This approach is similar to that devised by religious communities for their gods; "I refuse to prove I exist says God, for proof denies faith and without faith I am nothing". It is an untenable position.

The single biggest problem with QT (item 3 above) is that its atom can only absorb energy; it has no way to emit it. However, item 6 above also proves that Newton's is the correct atomic model.

QT was driven by a desire to explain events that were either unknown or misunderstood. Now that we fully understand the theory behind all orbital systems and that light does not possess mass, QT has become redundant, especially as it does not obey Newton's laws of orbital motion.

Whilst it is possible to create a sub-theory to explain any distortion of reality you wish, why would you if there is no need?

When everything in the universe can be explained without a sub-theory, the sub-theory becomes redundant.

It seems clear that Quantum Theory must be declared *'dead in the water'*, given that it fails to address the atom's single most important issue; the emission of electro-magnetic energy, whilst Newton's and Coulomb's laws together can explain all aspects of atomic structure and performance.

6.4 The Error

Relativity and quantum theory came about because of an error made prior to the 20th century. I believe that this error is responsible for having stalled scientific progress for a hundred years.

It is currently believed that electrons emit light (photons); **they don't**.

We have been taught this for a hundred years, forcing us to create fanciful theories to explain how mass moves in waves; **it doesn't**.

The photon exists in our minds because of a very simple mistake made a long time ago relating to Crooke's tube.

Crooke, and everyone since, believed that he had created a perfect vacuum by pumping out all the air from his tube; **he hadn't**! His tube contained a *measured* vacuum (refer to Chapter 6.4.1), it was not a *true* vacuum; there were millions of protons inside it. It is impossible to create a perfect vacuum on planet Earth, or anywhere on or in a celestial body where all matter resides.

When Crooke fired electrons from one end of his tube to the other, they appeared to emit light. So, he and everyone since believed that electrons must emit light. But the light you are seeing is not emitted by electrons, it is the electro-magnetic radiation emitted due to their interaction with protons in the tube. When a bar magnet is placed beside the tube, the light path deflects. What you see is not the bending of light (although magnetism does [slightly] bend light), the dramatic deflection you see is that of the *path* of the electrons; the light is emitted by interactions along this deflected path.

During his 'light-on-a-metal-plate' experiment Max Planck detected a feint but perceptible electric charge that '*pulsated*', confirming that it was induced by electro-magnetic energy. If it had been induced by a stream of electrons (photons) it would have been continuous as in a battery, confirming that light is electro-magnetic energy not electrons. The light emitted by stars and everything else in the universe is not brought to us by electrons: it is radiated electro-magnetic energy. Michael Faraday understood this and James Clerk Maxwell described it mathematically.

These are the main reasons why relativity and quantum theory remain unproven after 100 years; they both rely on light-emitting electrons.

6.4.1 Measured Vacuum

A *measured* vacuum defines the number of protons inside Crooke's Tube: Crooke would have had to remove every proton from inside his tube to claim that electrons fired within it were emitting light.

His tube was originally filled with air:
78% nitrogen, 21% oxygen & 1% argon.
Fig 49 shows the number of protons inside his tube at pressures between 7.5E-18 bar and 1E-05 bar.

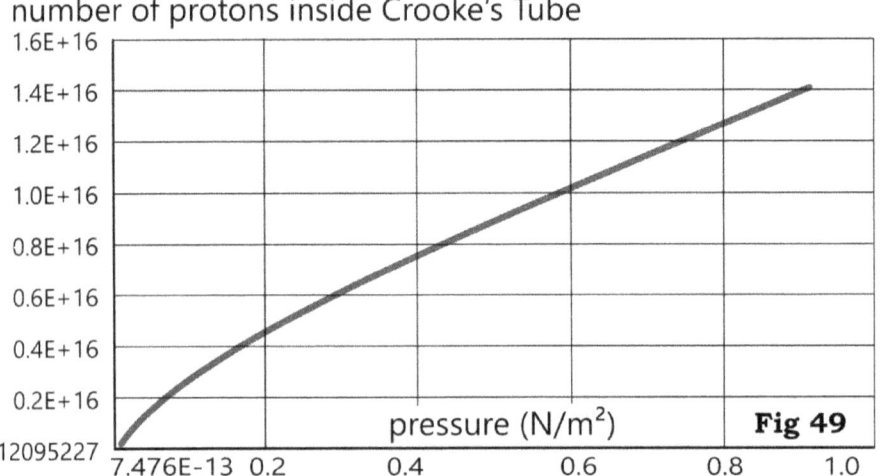

Fig 49

Gas pressure, including that in air, originates from the force generated between adjacent atoms by their electrical-charge repulsion.

This repulsion (force) drops off very quickly with the [square of the] distance between the atoms (exponential curve), and as can be seen from the above graph, even at 7.5E-13 N/m^2 (7.5E-018 bar), there would still have been about 12 million protons inside his tube.

Crooke used a mechanical pump to remove, what he thought was, all the air from inside it.

However, today's best laboratory mechanical pumps could achieve little better than 1E-10 bar (1E-05 N/mm^2), which relates to about 3E+12 protons left inside his tube. It is expected that Crooke actually achieved about 1E-05 bar, which meant there were about 1.4E+16 protons left inside his tube.

6.5 Model Verification

The following sub-chapters provide supporting mathematical verification of the atomic model described in this publication. But the relationship provided below is the final conclusive proof that the atomic model described in this book is correct:

Dalton's law states that each gas in a mixture of different gases will evenly fill its container independently of all the other gases in the mixture.

Partial pressure theory states that the total pressure of a gas mixture is the sum of the pressures of each individual gas.

Coulomb's force law between the electrical charges developed in adjacent protons with temperature is responsible creating gas pressure.

P.V = n.R_i.T:

Today, we calculate the pressure (p) of a gas in a container thus:
p = n.R_i.T / V
where: 'n' is the number of moles in the gas, 'R_i' is the ideal gas constant, 'T' is its temperature and 'V' its volume.

But it can also be calculated thus:
p = ρ.PE_1 / m_M.Y
where: 'ρ' is the gas density, 'PE_1' is the potential energy between the proton and the electron in shell-1 and 'm_M' is the molecular mass
PE = m_e.v_e^2
v_e = √[T/X]
which provides *exactly* the same result as the PVRT calculation method but is much simpler because there is no need to play with moles.

Because this latter calculation method replaces 'PVRT' but bypasses Boltzmann's constant, Avogadro's number, gas temperature and the ideal gas constant with potential energy and electron velocity, the model described in this book must be correct.

However, there are a few more proofs available ...

6.5.1 Density vs Temperature

The magnetic field energy (MFE) generated by proton-electron pairs holds adjacent atoms together (viscosity) and is constant irrespective of temperature. The electrical charges (EC) generated in protons repel adjacent atoms (gases) and varies between e and e' with temperature.

Given that temperature effects (e.g. gasification) are dependent upon repulsive EC between adjacent atoms and density is dependent upon an attractive MFE between those same atoms, and that both charges are created by the same energy generation process (proton-electron pairs), density and temperature should follow similar patterns of behaviour according to the number of nucleic protons (atomic number (Z)).

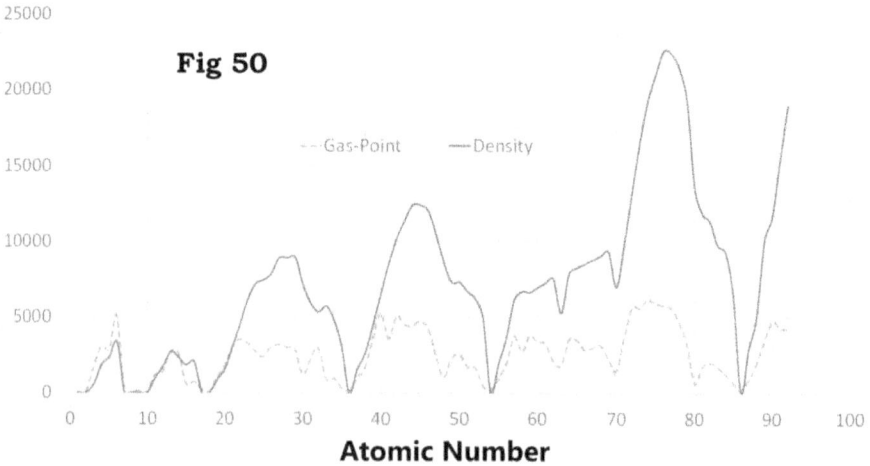

This relationship (between EC and MFE), which can clearly be seen in the temperature/density vs atomic number plot shown in Fig 50 is governed by the nucleic structure, which is the last significant piece of the atomic puzzle. Whilst it can be resolved mathematically (refer to Chapter 3.5.3), it has not been addressed here because it is not part of Isaac Newton's laws of orbital motion.

6.5.2 Specific Heat Capacity

The specific heat capacity of an atom defines the amount of energy it can hold in relation to its mass per unit temperature. This means the sum of the kinetic energy of all electrons in an atom's shells relative to its mass and *'temperature'*. The *'temperature'* in this case (as in all cases) is as defined in Chapter 2.1.6

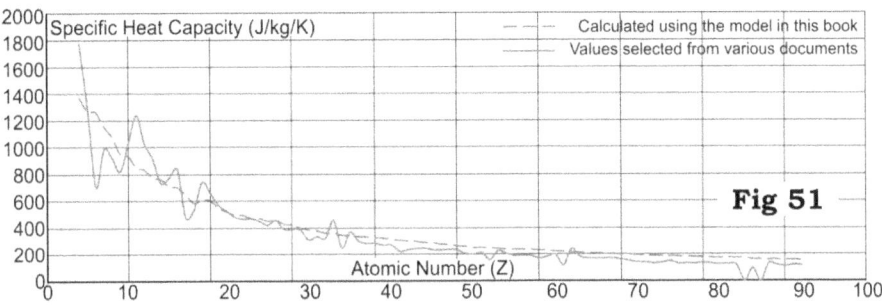

Fig 51

Fig 51 shows the calculated values for specific heat for all atoms from Z=4 to 92 compared to the documented values that have been taken from various sources and which are subject to experimental error.

This calculation technique, is as follows:

$SHC = KE_T / Y.m.T_1$ $\{J/kg/K\}$

where:
KE_T = the total kinetic energy in every electron in the elemental shells
T_1 = the temperature of the electron(s) in the innermost shell
Y (refer to Chapter 5.4)
m = the total mass of the atom (including electrons and neutrons)

6.5.3 Gas-Point

The gas-point of any atom is the temperature at which its electrical charge (EC) exceeds its magnetic field energy (MFE).

If the MFE is greater than the total exposed EC, the atoms will exist as viscous matter; otherwise they will exist as a gas.

Outlying nucleic neutrons protect adjacent atoms from EC. The more outlying neutrons, the greater the protection (higher gas-point temperatures and greater densities).

Density rises with atomic number (Fig 50) because larger atoms tend to collect a greater percentage of neutrons, due to the higher collective MFE within atoms.

Gas-point temperatures also rise with atomic number (Fig 52) for the same reason, but this rise is much less marked because only the outlying and exposed proton EC actively repels.

A mathematical relationship that reflects reality has been identified for all the atoms in the Periodic Table (refer to Chapter 3.5.3).

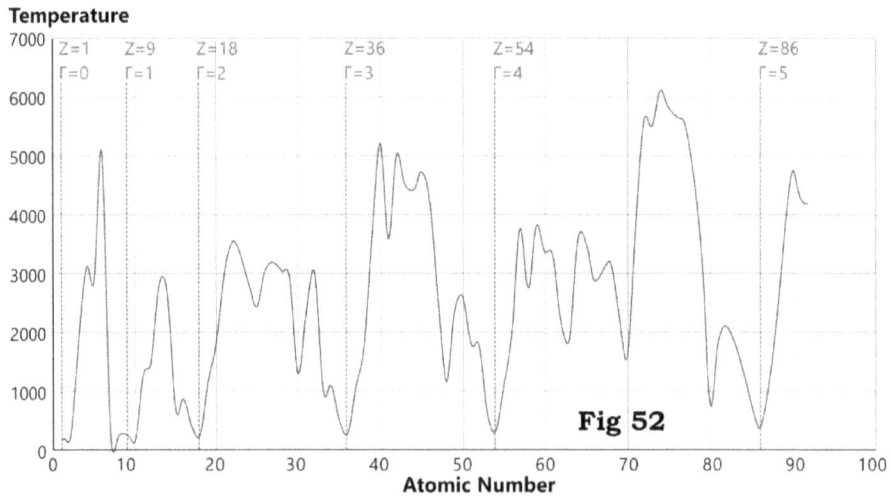

Fig 52

This relationship ($Z \to \Gamma$) forms the basis for mathematical chemistry.

6.5.4 Our Sun

Almost all (>99.7%) natural hydrogen exists as lone protons (H⁺), which cannot be solidified or liquefied due to their similar positive charges. Lone protons also have no way of absorbing or emitting electro-magnetic energy (heat and light).

It is currently claimed that our sun was created from a cloud of hydrogen atoms that accreted into a star due to gravity and an *external force*. This is of course, impossible (refer to Chapter 2.7.5)

The surface temperature of our sun is said to be about 5778K, which would be impossible if the sun's surface comprised lone protons that cannot collect or emit electro-magnetic energy (i.e. heat or colour).

Yet according to the atomic model proposed in this book, at 5778K;
KE = 3.7493802154296E-19 J
electron velocity = 912757.252 m/s
at an orbital radius of 3.03992067E-10 m
f = v / 2πR = 4.77873747733E+14 Hz
λ = c/f = **6.27346575162११E-07 m**

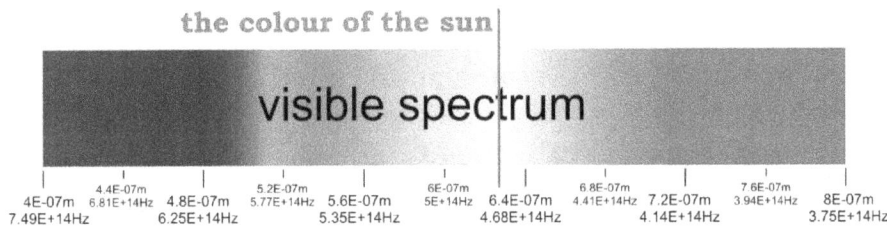

Fig 53

This demonstrates that the hydrogen at the surface of the sun is predominantly proton-electron pairs, which in turn means that these hydrogen atoms (proton-electron pairs) *must be a by-product of fission.*

6.5.5 PVRT

The final proof of this atomic model can be seen by the replacement of the well-known theory P.V = n.Ri.T with an alternative calculation using the potential energy in a proton-electron pair as described in Chapter 3.5.2

Today, we calculate the pressure (p) of a gas thus:
p = n.R$_i$.T / V
where: 'n' is the number of moles in the gas, 'R$_i$' is the ideal gas constant, 'T' is its temperature and 'V' its volume.

But it can also be calculated thus:
p = ρ.PE$_1$ / m$_M$.Y = T.m$_e$.ρ / X.Y.m$_M$ = k$_B$.T.ρ / m$_M$

where: 'ρ' is the gas density, 'PE$_1$' is the potential energy between the proton and the electron in shell-1 and 'm$_M$' is the molecular mass:
PE = m$_e$.v$_e^2$
v$_e$ = $\sqrt{[T/X]}$

which provides *exactly* the same result as the PVRT calculation method but is much simpler because there is no need to play with moles.

Because this latter calculation method replaces 'PVRT' altogether, along with the need for Boltzmann's constant, Avogadro's number, gas temperature and the ideal gas constant with potential energy, the model described here must be considered correct.

It is also interesting to note that the lattice structure we know applies to viscous matter (ζ : refer to Chapter 3.5.3) also applies to the same elements in gaseous form, and is responsible for *partial pressure*.

6.6 Heat

Heat is the electro-magnetic energy radiated by a proton-electron pair: the greater the energy, the greater the heat

Temperature is what we feel from the electro-magnetic energy radiated by a proton-electron pair with its electron(s) in the innermost shell: the greater the energy, the greater the temperature

We sense this heat and temperature through the energy in the electro-magnetism radiated by the proton-electron pairs.

Heat energy in an atom is the sum of the kinetic energies in all of its electrons.

The relationship between electron velocity (kinetic energy) and temperature may be defined as; $\underline{T} = X.v^2/e^2$ {K}
but as 'e' is a constant:
$\underline{T} = X.v^2$ {K}
Which is the same as: $\underline{T} = PE / k_B.^3\sqrt{[\,½ . \sqrt{[(4\pi)^2.a_o / R_n]}\,]}$
Where:
$X = \underline{T}.t_n^2/(2\pi.R_n)^2 = \underline{T}/c^2 = 6.9353271647894E\text{-}09$ K.s²/m²
and KE is the kinetic energy of the electron

The relationship between electron orbital radius and temperature may be defined as:
$\underline{T} = X_R/R$ {K}
Where: $X_R = \underline{T}_n.R_n = 1.75646616508036E\text{-}06$ {K.m}

You may have noticed that; $\underline{T} = X.v^2 / e^2$ is similar to Newton's gravitational force $F = G.m_1.m_2 / R^2$, Coulomb's force $F = k.Q_1.Q_2 / R^2$, and Gilbert's and Maxwell's formulas for force and energy. It is therefore anticipated that all of these formulas will eventually become just two; one for magnetic charge (gravity) and the other for electrical charge.

An interesting relationship for the above heat constants is as follows:
$4\pi^2.X/X_R = K$ {s²/m³}

Where K is Isaac Newton's orbital constant of proportionality for circular orbits (e.g. as in the atom):
$K = t^2/a^3 = 0.15587874533403$ {s²/m³}

All heat energy is radiated. Convection and conduction are simply different forms of radiation.

Conduction is simply the transfer of electro-magnetic radiation between proton-electron pairs within matter.

Convection is simply the balancing of electrical field energy between adjacent atoms. Atoms at a higher temperature (with greater heat energy) will try to move to a position where other atoms are further apart (e.g. cooler locations). In our atmosphere, this is always *upwards*, away from the earth's surface where the electrical repulsion energy (in a gas) can balance with the magnetic attraction energy (gravity).

6.7 Mass

I claim that there is no such thing as mass,
i.e. **mass** is magnetic charge, the magnitude of which is equal to the elementary charge unit (e):
$m = |e|$
$m' = |e'|$

Therefore; what is currently referred to as the elementary charge unit should be the *electrical charge unit* ($\pm e$), and the unit of mass should be the *magnetic charge unit* (m). The magnetic charge's non-polar nature is what causes all particles to attract all other particles.

Every particle holds a constant non-polar magnetic charge and also *retains the capacity* to hold the same magnitude of electrical charge.

For example;
The electron holds 'm' Coulombs of magnetic charge and '-e' Coulomb's of electrical charge constantly.
The proton holds 'm'' Coulombs of magnetic charge and '+e' Coulomb's of electrical charge. However, its greater magnetic charge, gives it the capacity to increase its electrical charge to '+e'' if and when partnered by an orbiting electron.

The number of particle assemblies (electron + proton + neutron) in a body:
$n = \text{mass} / 2.(m_e + m_p)$

The magnetic charge in each particle assembly:
$m = 2.(e+e') = 5.88688075484235\text{E-}16$ C

Using the planet Mercury to demonstrate this concept ...

First; we generate a new magnetic constant (M) based upon Coulomb's constant (k) as follows:

$M = c^2.m_e^2.R_n / m_p.e^2 = 4.89477777726655\text{E+}06 \text{ kg.m}^3 / \text{C}^2.\text{s}^2$

Note:
If $R_K = \sqrt{[G/M]} = 3.69243914581423\text{E-}09$ C/kg
$\{\sqrt{[(m^3 / s^2.kg)/(kg.m^3 / C^2.s^2)]} = C/kg\}$
and RC = 1.75881869180547E+11 C/kg
$R_K / RC = 2.09938589066497\text{E-}20 = \sqrt{\varphi}$
i.e. the square-root of the coupling ratio (refer to Chapter 6.11.3)

PHILOSOPHIÆ NATURALIS PRINCIPIA MATHEMATICA Revision IV

Next: we calculate the magnetic charge in Mercury and its force-centre:

	Mercury	**Sun**
mass (kg)	3.3011E+23	1.9885E+30
No. of particle groups	9.862670275E+49	5.941025671E+56
magnetic charge (C):	m_2 = 1.21891108642474E+15	m_1 = 7.34241524145157E+21
Table 6.7-1: Magnetic charge of the planet Mercury and its force-centre			

Finally: we use our new constant 'M' and the magnetic charges (m_1 & m_2) to calculate the 'gravitational' force between Mercury and the sun:

$F = M.m_1.m_2/R^2$ (R is the separation distance at Mercury's perigee)

F = **2.07016816968015E+22 N**
(refer to Chapter 4.2.2; F^P = **-2.0701682E+22 N**)
and it works for all the planets in our solar system

There is an additional argument:

Potential Energy:
The potential energy of every orbiting body may be calculated using the following formula:
PE = [($m_1.m_2$) ÷ (m_1+m_2)] . g.R {J}   **#1**
Where; g is the gravitational acceleration from m_1 at radius R
This formula (**#1**) produces exactly the same result as Isaac Newton's formula for the same property:
PE = $G.m_1.m_2/R$ {J}

PE = √ [($E_1.E_2$) ÷ (E_1+E_2)] . $g.R/v_c^2$ {J}   **#2**
E_1, E_2 & v_C : refer to Chapter 6.11.15

Both formulas (**#1** & **#2**) produce exactly the same potential energy between every force-centre and every satellite in our solar system confirming that mass may also be defined as magnetic energy.

Moreover, if 'n' represents the number of atomic particles in a celestial body and e is their Coulomb value; for every planet in the solar system:

g.R.e.($n_1.n_2$)/(n_1+n_2) = 2.08908021007445E-08 {$C.m^2/s^2$}

6.8 Gravity

Gravity is the potential energy in all matter due to the non-polar magnetism in atomic particles.

Non-Polar Magnetism is universal; it is always present in every atomic particle and it is all-pervasive; i.e. it radiates out in all directions and its strength does not diminish with distance. Whilst this form of magnetism is weak, it accrues collectively. I.e. the magnetic strength of 100 particles is 100 times stronger than that in one particle. This is why planets and stars - that comprise enormous numbers of particles - have such a strong magnetic attraction, but is not the case in, say, a cup.

Contrary to popular belief, this [non-polar] magnetic energy does not decrease with the square of the distance from its source. It only *appears* to do so because at increasingly larger radii the same force is distributed over a larger spherical area

Its effect is ever-present and constant throughout the universe, gradually pulling together all its matter, i.e. slowing down the expansion caused by the '*Big Bang*'. Eventually, all the particles in the universe will stop moving away from each other and gradually re-aggregate into another ultimate-body. The universe will then start all over again.

The concept *gravity is magnetism* changes nothing in terms of force and energy in the laws of orbital motion. Isaac Newton was correct. I.e. *exactly* the same results may be achieved either by using gravity (as Isaac Newton described it) or using magnetism (as William Gilbert and Hendrik Lorentz described it). The difference between gravity and magnetism is that we know exactly what magnetism is. We understand it and we can explain it in terms of energy. We don't, however, know what gravity is. Nobody, not even the great man himself, could explain what it is, what generates it and how it works.

All mass comprises a number of proton-electron pairs (neutrons are protons and electrons combined), each of which carries an elementary electrical charge (e). This electrical charge is responsible for both the electrical and the magnetic energy held by atomic particles.

Before explaining how the magnetic field is established, you need to know a couple of things:

'R_n' is the orbital radius of an electron travelling at the speed of light, when it combines with its proton to create a neutron. I.e. when the attractive magnetic energy exceed the repulsive electrical energy:
$R_n = RC.\mu_o.e / 4\pi = G.m_p / c^2.\varphi = 2.817937953839E\text{-}15 \text{ m}$

I.e. the magnetic constant defines the radius where magnetic attraction between an electron orbiting at the speed of light equals electrical repulsion.

Magnetic Field:
$B = 1/RC$ {kg/C}

Faraday's constant:
$F = N_A.e = 96485.3317942156$ {C/mol}

Relative atomic mass of an electron:
$RAM_e = RAM_H.m_e/m_p = 0.000548580318390698$ {g/mol}
relative atomic mass of a hydrogen atom:
$RAM_H = 1.00727638277233E\text{-}03$ {kg/mol}

Avogadro's number (constant):
$N_A = 6.02214129E+23$ {/mol}

Elementary charge:
$e = 1.60217648753E\text{-}19$ {C}

Electron mass: $m_e = 9.1093897E\text{-}31$ {kg}
Proton mass: $m_p = 1.67262163783E\text{-}27$ {kg}

Using the above information, it is possible to explain gravitational energy in terms of *non-polar* magnetic attraction. Lorentz gave us the following formula for magnetic attractive force:
$F = q.v.B$
Where: 'q' is the electrical charge, 'v' is the relative velocity and 'B' is the magnetic field. However, in our case, because the attracting masses are stationary, the relative velocity in this formula must be modified to use the gravitational acceleration between the two masses.

It can get a little tricky from here ...

PHILOSOPHIÆ NATURALIS PRINCIPIA MATHEMATICA Revision IV

Let me explain:

The magnetism referred to below is <u>non-polar</u> and all-pervasive. A polar [field] magnet is one that has all of its atomic nuclei aligned such that electro-magnetic energy is generated in one direction (e.g. north to south). Non-polar magnetism (charge) is always present in all atomic particles and acts in all directions. It is what holds the universe together, but it is much weaker than polar magnetism the amount by which it is weaker is called a coupling ratio: φ = 4.407421117923350E-40

Every mass (m) contains a specific number of particles (protons, electrons and neutrons). Because we already know that every particle possesses an electrical charge and that a neutron is a proton plus an electron, we can determine the number of electrical charges in any mass thus:
$q = m / (m_p + m_e)$
where m_p & m_e are the proton and electron masses respectively

Ok, here we go ...
Lorentz's magnetic force formula is;
$F = q.v.B$
where: q is the electrical charge; v is the velocity of the particle, and; B is the magnetic field.
His constant for the *magnetic field*:
$B = \mu_o.I / 2.\pi.R = 1/RC = 5.685634367312E-12$ {kg/C}
where: $R = 2.R_n$ & $I = e$
However, this approach is not very helpful for calculating the magnetic force between stationary particles.

<small>Note: B is currently referred to as the magnetic field. Whereas, it is actually the reciprocal of relative charge (B = 1/RC {kg/C})</small>

So we should use the following formulas for potential (rather than kinetic) force and energy:
$F = e.g/RC$ (kg.m/s²)
$E = e.g.R/RC$ (kg.m²/s²)

On the other hand, the following approach can be corroborated with Newton's and Coulomb's formulas for the magnetic potential energy between a proton and its orbiting electron:
$PE = e.v^2/RC \; (= R/RC)$ {J}
&
$F = PE/R$

Where: v is the velocity of the orbiting electron; RC is the relative atomic charge (B = 1/RC); g is the gravitational acceleration between the proton and the electron, and; R is the electron's orbital radius.
This is simply Lorentz's formula written in a useable format.

Because *exactly* the same magnetic forces and energies can be obtained universally for Newtons' gravitational calculations (in both astronomic and atomic environments), it is clear that gravity is actually magnetism.

Any two attracting masses (m_1 & m_2) may be described in terms of their static electrical charges (q_1 & q_2) which, in the case of an atom are equal to the elementary charge unit (e).

Exactly the same potential energy (PE) between m_p & m_e can be found using any and all of the following formulas:

Orbital: PE = $m_p.g$ (Isaac Newton)

Gravity: PE = $G/\varphi \cdot m_p.m_e/R$ (Isaac Newton)

Electrical: PE = $k.q_p.q_e/R$ (Coulomb)

Magnetic: PE = $q_p/RC \cdot g.R$ (Lorentz & Keith Dixon-Roche)

Heat: PE = $\underline{T}.k_B.m_p/m_e$ (Keith Dixon-Roche)

Where:
g is the gravitational acceleration between m_1 and m_2
R is radial separation between m_1 and m_2
B is the magnetic field (refer to Chapter 5.3: 1/RC)
k is Coulomb's constant
φ is the coupling ratio
m_p is the mass of a proton
m_e is the mass of an electron
q_p is the charge of a proton
q_e is the charge of an electron

The two important constants here are Isaac Newton's gravitational constant (G) and Coulomb's constant (k), both of which are based upon the properties of Quanta. It is important, therefore, to establish their relationship to each other.

PHILOSOPHIÆ NATURALIS PRINCIPIA MATHEMATICA Revision IV

$G = a_o.c^2 / \rho_u = 6.67359232004332\text{E-}11$ {m³ / kg.s² per m³}

$k = R_n.m_e.c^2/e^2 = 8.98755184732667\text{E+}09$ {kg.m³ / s².C²}
(refer to Chapter 6.11.4 and above)

so we can define the relationship as follows:
If: $Z \cdot a_o.c^2 / \rho_u = m_e.R_n.c^2/e^2$
Then: $Z = m_e.R_n.c^2.\rho_u / a_o.c^2.e^2$
{kg.m.m².kg.m³.s² / m³.m.m².s².C² = kg²/C²}

Newton's gravitational constant is also equal to:
$G = k.e^2.\varphi / m_e.m_p = 6.67359232004332\text{E-}11$ m³ / kg.s²
which applies to his formula: $F = G.m_1.m_2 / R^2$

After dividing out the mass components:
$m_1.m_2 \div m_e.m_p$, we get a product of particles '$n_1.n_2$'.
If G_e now $= k.e^2.\varphi = 1.01682605280249\text{E-}67$ kg.m³ / s²
And rewriting Newton's formula thus: $F = G_e.n_1.n_2 / R^2$
We get the same result, but it is now in terms of a number of elementary charge units

This calculation for the electrical potential energy between the sun and the earth at its perigee, gives: PE = 1.2208949335E+73 J, whereas the gravitational potential energy is: PE = 5.380981972219E+33 J, the difference between the two being 'φ'

Whilst magnetic energy is accrued, electrical energy is shared, locking the electrical attractive energy between an electron and its proton at the atomic level. This [electrical] energy does not pass beyond the atom. I.e. 'G_e' may only be used instead of 'G' in Isaac newton's formula for the atom. It cannot be used for the calculation of lunar, solar or galactic orbital systems.

The above conclusively demonstrates that Isaac Newton's laws of orbital motion apply equally well to atoms

It is no longer necessary to use the term gravity; Gravity is Magnetism.

Moreover, it conclusively unites Newton's gravitational laws of orbital motion with those for the atom, something that cannot, and no doubt never will be the case for quantum theory.

PHILOSOPHIÆ NATURALIS PRINCIPIA MATHEMATICA Revision IV

6.9 Electricity

What is electricity?

Atomic particles possess static, constant magnetic and electrical charges.

The movement of a static electrical charge relative to a static magnetic charge will generate magnetic field lines around the electrical charge. This phenomenon also works the other way around. Electrical field lines will similarly encircle a moving magnetic charge. These field lines follow the right-hand rule and we use this phenomenon to generate field electricity (AC).

If sufficient voltage is applied to both ends of an electrical conductor, dependent upon the number of coulombs in the conductor, it is possible to overcome the potential energy between the electrons in the conductor and their protons. When this happens, the electrons, being negatively charged will be attracted to the next available atomic shell valency in the positive direction of the conductor.

Voltage is generally written as PD (potential difference). Potential energy is the same thing, as can be seen from its energy-related units; J/C

By applying a potential energy across a conductor, you are attempting to overcome the PE in the proton-electron pairs in its atoms. With sufficient energy (per electron or Coulomb), you may be able to cause more than one or two electrons per atom to swap atomic orbits.

The ability to overcome atomic PE is also dependent upon the temperature of the conductor. As its temperature rises, PE between electrons and their protons increases, making it necessary to raise the energy (voltage) to overcome the proton-electron PE. This phenomenon is called electrical resistance.

The units for electricity (V, A, Ω, H, F, etc.) are useless when trying to calculate electrical energy. I have therefore described below all of these properties in terms of units we can use in tandem with mechanical and energy systems.

Given our new-found knowledge of the atom, we can test this theory:
1 Amp = 1 C/s
and
1 Volt is 1 J/C
and
1 Watt is 1 J/s
and
1 Ohm is 1 J.s/C²
and we know that each electron has a constant static charge of:
e = 1.60217648753E-19 C
We can use an example to demonstrate how it works.

A calculation for copper wire and a tungsten filament using Newton's and Coulomb's laws of orbital motion and are given below:

	Formula	Copper @ 273K	Tungsten @ 3000K	
Dia. (Ø)		0.002	0.000046	m
Length (L)		5	0.533	m
Density (ρ)		8960	19293	kg/m³
at mass	ρ.L.Π.Ø²/4	0.140743351	1.70896244E-05	kg
PE_n		-2.391841E-21	-1.064981E-20	J
n	$m / 2.m_n.Z$	1.7606391E+20	6.8998021E+19	at
PE	$n.PR_n$	-2.391841E-21	-114.23350	J
Power (W)		500		J/s
Volts (V)		250		J/C
Amps (A)	P/V	2		C/s
Flow-rate	A/e	1.24830193E+19		el/s
per atom	Flow/n	0.070900498	0.180918513	el/s/at
at = atom		m_n = mass of a neutron		
el = electron		PE_n = potential energy in outer shell		

The filament will consume 2 amps and the electron flow-rate will be 1 electron per atom every 5.5 seconds. Electrons flow from each atom in the copper wire every 14.1 seconds.

When an electron is pulled from its orbit, the potential energy between it and its proton (PE_n) is released. This is how electrical heat (electromagnetic energy) is generated (lost to the atom).

If the conductor temperature increases but the electrical power remains the same, PE_n will rise and the voltage necessary to transfer electrons will also rise, but the current that passes will fall (electrical resistance).

Because everything in the universe is electro-magnetic energy, our insistence on understanding everything in terms of units that currently do not apply (V, A, Ω, H, F, etc.) to electricity makes very little sense. I have therefore decided to describe the principal properties of electricity in the same units as everything else in this publication (refer to Chapter 5.8) Where:
Q = electro-magnetic radiation charge
f = electro-magnetic radiation frequency
E = electro-magnetic radiation energy
N_A = Avogadro's number
μ_o = magnetic constant (refer to Chapter 6.11.4)
R_n = neutronic radius (refer to Chapter 6.11.10)
m_e = mass of an electron
e = elementary charge unit

Current is the measure of electron flow rate along a conductor. So, this conversion is simple:
Amps: **A** = $e.f$ {**C/s**}

Voltage is the potential energy needed to generate current. This energy is shared between all electrons in the conductor and will only cause electrons to pull out of their proton-electron pair orbits (generate electricity) within the conductor if the energy is high enough. So, this conversion is also simple:
Volts: **V** = E/e {**J/C**}

Temperature: T = $2.E/k_B$ {J / J/K = K}
Where: E = electron kinetic energy

We know that **Resistance** is voltage divided by current:
Ω = V/A {J/C / C/s = **J.s/C²**}
But it can also be calculated in a few other ways:
Ω = ½.m_e.(v/e')² / f = π.k'/v = E / f.e'²
Where: v = electron velocity and E is electron kinetic energy

Power is Volts x Amps: P = C/s . J/C = J/s (Watts)

The **Farad** relates to the electrical charge capacity of matter, and we know that **electrostatic capacitance** is the electrical charge in an electron multiplied by the mole, so this conversion is also simple:
Farad: **F** = $e.N_A$ {**C/mol**}

It is also interesting to note that the Farad is the relative charge capacity of the electron, which can be demonstrated as follows:

RAC_e = 1F = $N_A.e$ = 6.02214129E+23 x 1.60217648753E-19
= 96485.3317942156 {C/mol}

This unit is the same as the electrical equivalent of RAM; *Relative Atomic Charge*, as it applies to the electron.

The RAC value for the proton is:
RAC_p = $N_A.e \cdot m_p/m_e$ = $N_A.e'$ = 177161652.983418 {C/mol}

The specific gas constant for the electron: R_a = R_i/RAC_e {J/C/K}
The specific gas constant for the proton: R_a = R_i/RAC_p {J/C/K}

The **Henry** relates to the electrical resistance of matter, and we know **mutual induction** (Henry) as the rate of current change induced in an electrical circuit, so the units must be C/s^2.

Given the units for magnetic constant; H/m:
μ_o = 4.π . 1E-07 H/m
1E-07 = $m_e.R_n/e^2$ kg.m/C^2
μ_o = 4.π.$R_n.m_e/e^2$ kg.m/C^2
Henry: **H** = $1/\mu_o$ {**kg.m^2/C^2**}

1 Henry is 1 Volt per Amp per second or 1 Coulomb per second squared
(V.t / A or C/s^2)
The Volt per Amp is electrical resistance (Ω = V/A)
Their units are: V {J/C} & A {C/s}
V/A {J.s/C^2}
H = V.t / A {J.s^2/C^2 = kg.m^2/C^2}
where 't' is the period of rate of change of current (as in acceleration)

6.10 Newton vs Planck

Newton's atom is the real one; the one we see, hear and feel all around us. Planck's atom is theoretical, it does not exist. It is based upon his three constants (Table 6.10.2 below); time (t^P), length (λ^P) and mass (m^P).
It should be noted here that Planck's length is actually an orbital radius (R) and not a wavelength (λ) as depicted below. The reason being; force (F) is Energy divided by length, and a length in such a formula can only be an orbital radius.

The atomic particles in Planck's atom are identical in size; his electron is exactly the same, in all respects, as his proton. The only consistency between the two atoms (Newton's and Planck's) is that the product of the particle volumes in each atom is identical; $V_e.V_p = V^P.V^P = 3.0\text{E-}91 \text{ m}^6$ (Refer to Chapter 6.11.1)

It is the comparison between these two atoms that has given us the ability to solve Newton's gravitational constant (G; refer to Chapter 6.11.2)

The following Tables (6.10A to 6.10C) provide comparative results for three atomic variations in which the particle properties are defined thus:

Newton atom; comprising Newtonian particles (Table 6.10A)
$t = a_o/c$; $\lambda = a_o$; $m = \rho_u$; $E = \rho_u.c^2$; $F = E/a_o$

Planck^N atom; comprising Newtonian particles (Table 6.10A)
t^P; λ^P; m^P; E^P; F^P (Table 6.10B)

Planck^P atom; comprising Planck particles (Table 6.10C), and calculated using a revised version of his own constant; $\hbar$:
$\hbar = \sqrt{[\pi.m.e^2.a_o / \varepsilon_o]} = 7.99473592559180\text{E-}16 \text{ kg.m}^2/\text{s}$
in which:
$a_o = \lambda^P / 4\pi^2$
$e = \sqrt{[G.m^{P2} / k.\varphi]}$
$m = m^P$

Note: 'c', 'ε_o', 'G', 'φ' and 'k' are equal in both Newton and Planck atoms

PHILOSOPHIÆ NATURALIS PRINCIPIA MATHEMATICA Revision IV

Newton atom compared with a Planck^N atom:

	Newton Atom (A)	Planck^N Atom (B)	A/B
t	1.76514516887831E-19	5.39096122598359E-44	3.27426797353056E+24
λ	5.2917721067E-11	1.61616952231128E-35	3.27426797353056E+24
m	7.1266079635045E+16	2.1765500017459E-08	3.27426797353056E+24
E	6.40507585675678E+33	1.95618559889902E+09	3.27426797353056E+24
F	1.21038391820525E+44	1.21038391820525E+44	1

Table 6.10A

Newton atom compared with a Planck^P atom:

	Newton Atom (A)	Planck^P Atom (B)	A/B
t	1.76514516887831E-19	1.48432887846076E-34	1.18918737922067E+15
λ	5.2917721067E-11	4.44990604438464E-26	1.18918737922067E+15
m	7.1266079635045E+16	59.9283854507006	1.18918737922067E+15
E	6.40507585675678E+33	5.38609471364748E+18	1.18918737922067E+15
F	1.21038391820525E+44	1.21038391820525E+44	1

Table 6.10B

Planck^N atom compared with a Planck^P atom:

	Planck^N Atom (A)	Planck^P Atom (B)	A/B
t	5.39096122598359E-44	1.48432887846076E-34	2.75336589568949E+09
λ	1.61616952231128E-35	4.44990604438464E-26	2.75336589568949E+09
m	2.1765500017459E-08	59.9283854507006	2.75336589568949E+09
E	1.95618559889902E+09	5.38609471364748E+18	2.75336589568949E+09
F	1.21038391820525E+44	1.21038391820525E+44	1

Table 6.10C

Note: 3.27426797353056E+24 ÷ 1.18918737922067E+15 = 2.75336589568949E+09

6.10.1 Newton's Atom (particle properties)

The following Table provides the properties of particles in the atom that exists in nature.

Symbol	Formula	Value	Units
m_e			
The mass of an electron (Table 2)			
m_p			
The mass of a proton (Table 2)			
m_n	$m_e + m_p$	1.6735325768E-27	kg
The mass of a neutron [4]			
V_e	m_e / ρ_u	1.27822236702922E-47	m^3
The volume of an electron			
V_p	m_p / ρ_u	2.34700946985653E-44	m^3
The volume of an proton			
V_n	m_n / ρ_u	2.34828769222356E-44	m^3
The volume of an neutron			
r_e	$\sqrt[3]{3 \cdot V_e / 4\pi}$	1.45046059426276E-16	m
The radius of an electron			
r_p	$\sqrt[3]{3 \cdot V_p / 4\pi}$	1.77613270336827E-15	m
The radius of a proton			
r_n	$\sqrt[3]{3 \cdot V_n / 4\pi}$	1.77645508248591E-15	m
The radius of a neutron			
t^N	a_o / c	1.765145168878310E-19	s
Newton time			
λ^N	a_o	5.291772106700000E-11	m
Newton length			
m^N	$a_o \cdot c^2 / G$	7.126607963504500E+16	kg
Newton mass (refer to Chapter 6.7)			
E^P	$m^N \cdot c^2$	6.405075856756780E+33	J
Newton energy			
F^N	E^P / λ^N	1.210383918205250E+44	N
Newton force			
ρ_u	$\sqrt{[m_e \cdot m_p]} / \sqrt{\Sigma}$	7.1266079635045E+16	kg/m^3
The ultimate density of Newton's atomic particles			

Table 6.10.1

6.10.2 Planck's Atom (particle properties)

The following Table provides the properties of particles in a fictitious atom that has been constructed using Planck's values; t^P, λ^P, m^P.

Symbol	Formula	Value	Units
m_e^P	m^P	2.1765500017459E-08	kg
Planck's electron mass, which is equal to Planck's mass			
m_p^P	m^P	2.1765500017459E-08	kg
Planck's proton mass, which is equal to Planck's mass (and Planck's electron)			
m_n^P	N/A	N/A	N/A
Planck's neutron is unnecessary as his atom is theoretical only			
V_e^P	$\sqrt{\Sigma}$	5.47722557505166E-46	m³
Planck's electron volume			
V_p^P	$\sqrt{\Sigma}$	5.47722557505166E-46	m³
Planck's proton volume			
V_n^P	N/A	N/A	N/A
Planck's neutron is unnecessary as his atom is theoretical only			
r_e^P	$\sqrt{[\,3.V_e^P\,/\,4\pi\,]}$	5.07563837996471E-16	m
Planck's electron radius			
r_p^P	$\sqrt{[\,3.V_p^P\,/\,4\pi\,]}$	5.07563837996471E-16	m
Planck's proton radius			
r_n^P	N/A	N/A	N/A
Planck's neutron is unnecessary as his atom is theoretical only			
t^P	$\sqrt{[\,\hbar.G\,/\,c^5\,]}$	5.39096122598359E-44	s
Planck time			
λ^P	$\sqrt{[\,\hbar.G\,/\,c^3\,]}$	1.61616952231128E-35	m
Planck length			
m^P	$\sqrt{[\,\hbar.c\,/\,G\,]}$	2.1765500017459E-08	kg
Planck mass (refer to Chapter 6.7)			
E^P	$m^P.c^2$	1.95618559889902E+09	J
Planck energy			
F^P	$C^8\,/\,G^2$	1.21038391820525E+44	N
Planck force			
ρ_u^P	$m^P\,/\,V_e^P$	3.97381844498046E+37	kg/m³
The ultimate density of Planck's atomic particles			

Table 6.10.2

6.11 Important Constants (explained)

It is a fact that many natural constants are incorrect, approximate or misunderstood. It is the aim of this Chapter to finally resolve this issue.

Each of the following sub-Chapters describe the reasoning behind the units and values given for the most important physical constants you will find in the main body of the text.

6.11.1 Σ

This universal constant seems to tie everything together including Newton's and Planck's particles: Σ = 3E-91 (exact). Its importance is self-evident as it also gives us 'ρ_u' and 'G'. It has such a bizarrely accurate value, however, that it may or may not exist (see below). The trouble is, it seems to appear everywhere. E.g.:

$G = \sqrt{[\Sigma.a_o^2.c^4 / m_p.m_e]}$

$m_e.m_p = \Sigma.\rho_u^2$

$V_e.V_p = \Sigma$
$V^{P2} = r^{P6}.(4/3.\pi)^2 = \Sigma$

$r^P = \sqrt[6]{[\Sigma / (4/3.\pi)^2]} = 5.07563837996471E-16$ m
is the radius of a Planck particle

If $F^N = G.m_e.m_p / a_o^2$ then; $F^N/F^P = \Sigma$
(refer to Chapter 6.10; Tables A to C; F^P)

Its units *appear* to be m⁶. I say *"appears"* because it also unites the Newton and Planck forces (F^N & F^P; see above), which means it can also have no units.

Σ defines the ratio; "*electron orbital radius*" : "*particle centres*" (ϑ)
That is; the relative radii of the orbiting electron and the "electron radius plus proton radius" (ϑ = R / [r_e+r_p])
For example; if Σ = 3E-91, at a velocity of 'c'; ϑ = 1.46677550700175
The following Table shows the changes that would result in modifying Σ to give us ϑ = 1.0 (i.e. the particles touch at 'c').
All other physical constants unchanged.

Σ	3E-91	2.98746723133494E-90	{m⁶}
ϑ	1.46677550700175	1.0	
G	6.67359232004334E-11	2.10596243650527E-10	m³ / kg.s²
ρ_u	7.12660796350450E+16	2.25835348953929E+16	kg/m³
φ	4.407421117923340E-40	1.39083463164654E-39	
r_e	1.45046059426276E-16	2.12750007353581E-16	m
r_p	1.77613270336827E-15	2.60518794648537E-15	m

Whilst; Σ = 3E-91 is not (yet) cast in stone, it does provide the most acceptable results for particle density, which is the basis for light-deflection and gives the correct value for G (refer to Chapter 6.11.2)

6.11.2 G

Today, you will see the units for this constant written as: $N.m^2/kg^2$, which were units of convenience originally assigned to reflect Newton's formula:
$F = G.m_1.m_2 / R^2$
This was because the formula for 'G' was unknown, hence its units were unknown.

From the relationship between Newton's Atom and Planck's Atom (refer to Chapter 6.11.1):
$F^N/F^P = \Sigma$ (no units)
where:
$F^N = G.m_p.m_e/a_o^2$ {N}
$F^P = c^4/G$ {N}
$G = \sqrt{[\Sigma.a_o^2.c^4 / m_p.m_e]}$
$= 6.67359232004334E\text{-}11$ $\{\sqrt{[m^2.m^4 / s^4.kg^2]} = m^3 / s^2.kg\}$
giving us its value and its units.

From Newton's laws of orbital motion, we know that a force-centre mass may be found thus: $m_1 = (2\pi)^2 / G.K$, and from Chapter 6.11.14 of this book, we know that: $K = (2\pi/v)^2 / R$
Therefore: $G = v^2.R/m_1$ confirming the units for 'G' as; $m^3 / kg.s^2$

If, on the other hand, Σ has units; m^6
$V_p = m_p/\rho_u$
$V_e = m_e/\rho_u$
$\Sigma = V_p.V_e = m_p.m_e/\rho_u^2$
$G.m_p.m_e/a_o^2 \div c^4/G = m_p.m_e/\rho_u^2$
$G^2 / a_o^2.c^4 = 1/\rho_u^2$
then;
$G = a_o.c^2/\rho_u$ $\{m^6/kg/s^2\}$
Because we know its value and units, 'G' must be a coefficient per unit volume ($m = V.\rho$) giving us:
i.e. $G = a_o.c^2 / m_u$ $\{m^3/kg/s^2\}$ per m^3
$G = a_o.c^2 / m_u$
$G = 5.2917721067E\text{-}11 \times 299792459^2 / 7.12660796350450E\text{+}16$
$= 6.67359232004334E\text{-}11$ $m^3 / s^2.kg$ per m^3

If, therefore, G = gravitational acceleration multiplied by spherical area per unit mass, the following calculations show that, contrary to popular belief, gravitational force does not vary with the square of the distance from the centre of its source.

PHILOSOPHIÆ NATURALIS PRINCIPIA MATHEMATICA Revision IV

Factors **1.5** & **4** below are exact values and will be explained in due course

*Note: A = **1.5**/a_o*

$G = \mathbf{1\cdot5}.c^2 / A.\rho_u = 6.67359232004334E-11$
$\{m^2/s^2 / (m^2.kg/m^3) = m^3 / s^2.kg\}$
$A = \mathbf{1\cdot5}.c^2 / G.\rho_u = 2.83458918818674E+10$
$\{m^2/s^2 / (m^3 / s^2.kg \cdot kg/m^3) = m^2\}$
$R = \sqrt{[A / 4.\pi]} = 47494.1512680647 \{m\}$ #
$V = {}^4/_3.\pi.R^3 = 4.48754692288540E+14 \{m^3\}$
$R_s = 2.G.m / c^2 = 47494.1512680647 \{m^3 / s^2.kg \cdot kg / (m^2/s^2) = m\}$ #
$R = R_s$
$m = R.c^2 / 2.G = 3.19809876372352E+31 \{m \cdot m^2/s^2 / (m^3 / s^2.kg) = kg\}$
$\rho = m/V = 7.12660796350450E+16 \{kg/m^3\}$
$\rho = \rho_u$

$G = R_s.c^2 / 2.m = 6.67359232004334E-11 \{m \cdot m^2/s^2 / kg = m^3 / s^2.kg\}$
$R_s = 2.G.m / c^2 = 47494.1512680647 \{m^3 / s^2.kg \cdot kg / (m^2/s^2) = m\}$ #
$g = G.m / R^2 = 9.46174592804013E+11 \{m^3 / s^2.kg \cdot kg / m^2 = m/s^2\}$
$F^N = G.m^2 / R^2 = 3.02595979551312E+43$
$\{m^3 / s^2.kg \cdot kg^2 / m^2 = kg.m/s^2 = N\}$
$F^P = c^4 / G = 1.21038391820525E+44$
$\{m^4/s^4 / (m^3 / s^2.kg) = kg.m/s^2 = N\}$
$F^P/F^N = \mathbf{4}$
$R_s = \sqrt[4]{[F^P / (^4/_3.\pi)^2.G.\rho^2.\mathbf{4}]} = 47494.1512680647$ #
$\{\sqrt[4]{[kg.m/s^2 / (m^3 / s^2.kg) / (kg^2/m^6)]} = m\}$
Schwarzschild radius (R_s) of mass 'm'

From the above formulas:
$G = g.A / 4.\pi.m = 6.67359232004334E-11 \{m^3 / s^2.kg\}$
$4.\pi.G = g.A / m = 8.38628344228057E-10 \{m^3 / s^2.kg\}$
Indicating that $F^N = 4.\pi.G.m_1.m_2 / 4.\pi.R^2$

In other words; Newton's gravitational force equation should read:
$F^N = G.m_1.m_2 / A$
and his gravitational constant should be:
$G = 8.38628344228057E-10 \text{ m}^3/\text{s}^2/\text{kg}$;
i.e. gravitational force is constant irrespective of distance from the centre of its source, but it will vary at distance (R) according to its distribution over the spherical area, upholding the conservation of energy law.

Two aspects of this discovery require further explanation:
1) $a_o = \mathbf{1 \cdot 5}$ / A (A = 2.83458918818674E+10 m²); what does 'A' represent?
2) $F^P/F^N = \mathbf{4}$ (exact), which also applies to Newton's formula for the deflection of light (refer to Chapter 6.2.1) and surface area A = $\mathbf{4}.\Pi.R^2$ of a sphere over which Newton's and Coulomb's forces are spread.

Both of the above will be addressed in a later edition of this publication but neither of which alters the final result.

And finally: according to Planck: G = $2\pi.\lambda^2.c^3$ / h = $\lambda^2.c^3$ / $\hbar$ {m³ / s².kg} confirming the units for 'G'.

In this solution for 'G', Planck's atom and Newton's theories have been fully analysed and complement each other perfectly. The discovery of so many atomic associations with G means that Newton actually did anticipate both Poincaré and Planck, confirming that his theories can be applied throughout all science, from the largest to the smallest.

i.e. there is no need for a unification theory (*Quantum theory is dead*)

6.11.3 φ

The coupling ratio is the ratio between gravitational (magnetic) charge (E_g) and electrical charge (E_e) and is defined as follows:
$E_g = G.m_1.m_2 / R$ and $E_e = k.q_1.q_2 / R$ {J}

$\varphi = E_g / E_e$
$\varphi = G.m_1.m_2 / k.q_1.q_2$

$\varphi = \dfrac{(6.67359232004334\text{E-}11 \times 9.1093897\text{E-}31 \times 1.67262163783\text{E-}27)}{(8.98755184732667\text{E+}09 \times 1.60217648753\text{E-}19 \times 1.60217648753\text{E-}19)}$

$\varphi = 4.40742111792334\text{E-}40$

This means that gravitational energy can only alter the density of matter if there is sufficient mass to generate the required pressure ($1/\varphi$) and this only occurs in the ultimate-body

The following relationship is also true:

$\varphi = V_p.a_o / R_n$

V_p; Refer to Chapter 6.10.1
a_o; Refer to Chapter 5.4
R_n; Refer to Chapter 5.3

6.11.4 k, k', μ_o, ε_o

Coulomb's constant is equivalent to Isaac Newton's gravitational constant (G) when applied to *electrical* force in exactly the same fashion. I.e.:

Isaac Newton's formula for gravitational force:
$F = G.m_1.m_2 / R^2$

Charles-Augustin de Coulomb's formula for electrical force:
$F = k.q_1.q_2 / R^2$

The difference between the two is defined as the coupling ratio:
$\varphi = G.m_p.m_e / k.q_p.q_e = 4.40742111792335E-40$

Given the following ...
... for the magnetic constant:
$\mu_o = 4\pi.\mu'$ H/m
$\mu' = 1.0E-7 = \mu_o / 4\pi$ {H/m}
$\mu' = 1.0E-7 = m_e.R_n / e^2$ {kg.m/C²}
$\mu_o = 4\pi.m_e.R_n / e^2$ {kg.m/C²}
Therefore, the unit 'Henry' is actually; kg.m²/C²

... and permittivity in a vacuum:
$\varepsilon_o = 1 / \mu_o.c^2$ {C².s² / kg.m³}

Coulomb's constant (k) may be calculated as follows:
$k = \mu_o.c^2 / 4\pi = 8.98755184732667E+09$
or
$k = \mu'.c^2 = m_e.R_n.(c/e)^2 = 1 / 4\pi.\varepsilon_o = \mu_o.c^2 / 4\pi$
$k = 8.98755184732667E+09$ {kg.m³ / C².s²}
When modified for electro-magnetic emission via protons:
$k' = k/\xi_m^2$

It is also interesting to note that the fine-structure constant is equal to:
$a = e^2 / 4\pi = 2.h'.\varepsilon_o = 2.0427294212227E-39$ {C²}

And the modified version of Planck's constant (refer to Chapter 6.11.5; h') is related to Coulomb's constant thus:
$h' = e^2 / 8\pi.\varepsilon_o = \frac{1}{2}.k.e^2 = 1.15353857232684E-28$ {kg.m³/s²}

Moreover; electrical resistance x electron velocity is a constant:
311971413.341244 J.m/C² (irrespective of temperature)
Whereas Coulomb's constant k = 8.98755184732667E+09 J.m/C²

6.11.5 h, h'

Because we know that energy cannot be created, it can only be transferred, the electro-magnetic energy emitted by a proton-electron pair must be the same as the kinetic energy in the orbiting electron. Moreover, the frequency (f) and amplitude (A) of electro-magnetic radiation is also equal to that of the orbiting electron (A = R and f = v / 2πR).

Max Planck claimed that the energy of electro-magnetic radiation can be calculated using his constant as follows:
E^P = h.f
in which his constant (h) is defined thus:
h = $\sqrt{[π.m_e.e^2.a_o / ε_o]}$ = 6.62607174469163E-34 **J.s**
However, these units (J.s) can only be considered correct if a frequency ratio is applied to his constant thus:
h = $\sqrt{[π.m_e.e^2.a_o / ε_o]}$. f_1/f_2
In reality, the units for Planck's constant should be:
$\sqrt{[kg.C^2.m / (C^2/m/(kg.m^2/s^2)]}$ = $\sqrt{[kg^2.m^4/s^2]}$ = **kg.m²/s**

h = $\sqrt{[π.m_e.e^2.a_o.4π.m_e.R_n.c^2 / e^2]}$
 = $\sqrt{[4π^2.m_e^2.c^2.a_o.R_n]}$
h = $\sqrt{[(4π)^2.a_o . R_n]}$. ½m_e.c {identical units; kg.m²/s}

Planck therefore *actually* identified a range of orbital radii:
a maximum: (4π)².a_o
a minimum: R_n
and a mean: $\sqrt{[(4π)^2.a_o . R_n]}$
The maximum orbital velocity: c

Because we know Planck's mean orbital radius: R_m = 2π.$\sqrt{[(4π)^2.a_o . R_n]}$
R_m = 4.852618433622630E-12 m
and we also know the mean velocity: v_m = $\sqrt{[X_R / X.R_m]}$
v_m = 7224342.80705001 m/s
and we can calculate his minimum electron velocity (v_o) from:
v_o = c . $\sqrt{[R_n / (4π)^2.a_o]}$ = 174090.866621082 m/s

However, if we modify Planck's constant thus:
h' = h.v_o = 1.15353857232684E-28 **kg.m²/s . m/s**
his modified formula becomes:
h' = $\sqrt{[(4π)^2.a_o . R_n]}$. ½.m_e . v_o.c = ½.R_n.m_e.c^2 {kg.m³ / s² = J.m}
and because we can also calculate temperature (refer to Chapter 6.6), we can find everything we need to know about his orbiting electron:

Energy:	min	mean	max
R (m)	8.35643156381572E-09	4.85261843362268E-12	2.817937953839E-15 #
v (m/s)	174090.866621084	7224342.80705001	299792459
KE (J)	1.38042005551962E-20	2.37714666443634E-17	4.09355561131261E-14
PE (J)	-2.76084011103925E-20	-4.75429332887267E-17	-8.18711122262522E-14
T (K)	210.193328535837	361962.554671561	623316124.717179

Table 6.11.5-1: Planck's Electron Performance Range

Note: # the neutronic radius (R_n) is 2.81793795383896E-15 m

Whilst Max Planck's original formula for electro-magnetic energy ($E^p = h.f$) is incorrect, an alternative calculation method ($E = h'/A$) using the modified version of his constant (h') does work (refer to Table 6.11.5-2 below).

Refer to Chapter 4.1.3 for the properties of electro-magnetic radiation associated with the above temperatures.

Another interesting relationship with h':
e.k/h' = 2   {$C^2 \cdot kg.m^3 / C^2.s^2 \cdot s^2 / kg.m^3$ = no units}

Moreover; if we extract $\sqrt{[(4\pi)^2 \cdot a_o \cdot R_n]}$ from **h** and modify it slightly, thus:
$\sqrt{[(4\pi)^2 \cdot a_o / R_n]}$ = 1722.0458764934
we get the velocity ratio ξ_v (refer to Chapter 5.3)

So, whilst Max Planck's claim regarding electro-magnetic energy using his constant was erroneous, without his work and proposal(s) these solutions would have been much more difficult to achieve; thank you Max!
h' = h.v$_o$ = ½.k.e² = e² / 8.π.ε$_o$ = ½.R$_n$.m$_e$.c² {kg.m³/s²}

In fact, Max Planck is the *only* 20th Century scientist without whom none of these scientific discoveries would have been possible.

PHILOSOPHIÆ NATURALIS PRINCIPIA MATHEMATICA Revision IV

T	R (A)	v	KE	E^P (error)	E (error)
6.2332E+08	2.8179E-15	299792459	*4.0936E-14*	1.12193E-11 (274.1)	4.09356E-14 (1)
3.1166E+08	5.6359E-15	211985280.7	*2.0468E-14*	3.96662E-12 (193.8)	2.04678E-14 (1)
2.0777E+08	8.4538E-15	173085256.9	*1.3645E-14*	2.15915E-12 (158.2)	1.36452E-14 (1)
1.5583E+08	1.1272E-14	149896229.5	*1.0234E-14*	1.40241E-12 (137.0)	1.02339E-14 (1)
1.2466E+08	1.4090E-14	134071263.5	*8.1871E-15*	1.00348E-12 (122.6)	8.18711E-15 (1)
1.0389E+08	1.6908E-14	122389758.9	*6.8226E-15*	7.63376E-13 (111.9)	6.82259E-15 (1)
8.9045E+07	1.9726E-14	113310898.8	*5.8479E-15*	6.05785E-13 (103.6)	5.84794E-15 (1)
7.7915E+07	2.2544E-14	105992640.4	*5.1170E-15*	4.95828E-13 (96.90)	5.11694E-15 (1)
6.9257E+07	2.5361E-14	99930819.7	*4.5484E-15*	4.15529E-13 (91.36)	4.54840E-15 (1)
6.2332E+07	2.8179E-14	94802699.6	*4.0936E-15*	3.54785E-13 (86.67)	4.09356E-15 (1)
5.6665E+07	3.0997E-14	90390827.4	*3.7214E-15*	3.07522E-13 (82.64)	3.72141E-15 (1)
5.1943E+07	3.3815E-14	86542628.5	*3.4113E-15*	2.69894E-13 (79.12)	3.41130E-15 (1)
4.7947E+07	3.6633E-14	83147467.9	*3.1489E-15*	2.39359E-13 (76.01)	3.14889E-15 (1)
4.4523E+07	3.9451E-14	80122904.9	*2.9240E-15*	2.14177E-13 (73.25)	2.92397E-15 (1)
4.1554E+07	4.2269E-14	77406080.1	*2.7290E-15*	1.93121E-13 (70.77)	2.72904E-15 (1)
3.8957E+07	4.5087E-14	74948114.7	*2.5585E-15*	1.75302E-13 (68.52)	2.55847E-15 (1)
3.6666E+07	4.7905E-14	72710351.4	*2.4080E-15*	1.60064E-13 (66.47)	2.40797E-15 (1)
3.4629E+07	5.0723E-14	70661760.2	*2.2742E-15*	1.46912E-13 (64.60)	2.27420E-15 (1)
3.2806E+07	5.3541E-14	68777107	*2.1545E-15*	1.35468E-13 (62.88)	2.15450E-15 (1)
3.1166E+07	5.6359E-14	67035631.7	*2.0468E-15*	1.25436E-13 (61.28)	2.04678E-15 (1)
2.9682E+07	5.9177E-14	65420077.9	*1.9493E-15*	1.16583E-13 (59.81)	1.94931E-15 (1)
2.8333E+07	6.1995E-14	63915967	*1.8607E-15*	1.08726E-13 (58.43)	1.86071E-15 (1)
2.7101E+07	6.4813E-14	62511048.9	*1.7798E-15*	1.01712E-13 (57.15)	1.77981E-15 (1)
2.5972E+07	6.7631E-14	61194879.4	*1.7057E-15*	9.54220E-14 (55.94)	1.70565E-15 (1)
2.4933E+07	7.0448E-14	59958491.8	*1.6374E-15*	8.97544E-14 (54.81)	1.63742E-15 (1)
2.3974E+07	7.3266E-14	58794138.4	*1.5744E-15*	8.46263E-14 (53.75)	1.57444E-15 (1)
2.3086E+07	7.6084E-14	57695085.6	*1.5161E-15*	7.99687E-14 (52.75)	1.51613E-15 (1)
2.2261E+07	7.8902E-14	56655449.4	*1.4620E-15*	7.57231E-14 (51.79)	1.46198E-15 (1)
2.1494E+07	8.1720E-14	55670062.1	*1.4116E-15*	7.18404E-14 (50.89)	1.41157E-15 (1)
2.0777E+07	8.4538E-14	54734364.1	*1.3645E-15*	6.82785E-14 (50.04)	1.36452E-15 (1)
2.0107E+07	8.7356E-14	53844315.1	*1.3205E-15*	6.50014E-14 (49.22)	1.32050E-15 (1)
1.9479E+07	9.0174E-14	52996320.2	*1.2792E-15*	6.19784E-14 (48.45)	1.27924E-15 (1)
1.8888E+07	9.2992E-14	52187168.5	*1.2405E-15*	5.91827E-14 (47.71)	1.24047E-15 (1)
1.8333E+07	9.5810E-14	51413982.6	*1.2040E-15*	5.65910E-14 (47.00)	1.20399E-15 (1)
1.7809E+07	9.8628E-14	50674174.5	*1.1696E-15*	5.41831E-14 (46.33)	1.16959E-15 (1)
1.7314E+07	1.0145E-13	49965409.8	*1.1371E-15*	5.19412E-14 (45.68)	1.13710E-15 (1)
1.6846E+07	1.0426E-13	49285576.7	*1.1064E-15*	4.98498E-14 (45.06)	1.10637E-15 (1)
1.6403E+07	1.0708E-13	48632758.7	*1.0773E-15*	4.78950E-14 (44.46)	1.07725E-15 (1)
1.5982E+07	1.0990E-13	48005213	*1.0496E-15*	4.60647E-14 (43.89)	1.04963E-15 (1)
1.5583E+07	1.1272E-13	47401349.8	*1.0234E-15*	4.43482E-14 (43.33)	1.02339E-15 (1)
1.5203E+07	1.1554E-13	46819716.1	*9.9843E-16*	4.27356E-14 (42.80)	9.98428E-16 (1)
1.4841E+07	1.1835E-13	46258980.7	*9.7466E-16*	4.12185E-14 (42.29)	9.74656E-16 (1)
1.4496E+07	1.2117E-13	45717921.4	*9.5199E-16*	3.97810E-14 (41.80)	9.51990E-16 (1)
1.4166E+07	1.2399E-13	45195413.7	*9.3035E-16*	3.84403E-14 (41.32)	9.30354E-16 (1)
1.3851E+07	1.2681E-13	44690421.2	*9.0968E-16*	3.71661E-14 (40.86)	9.09679E-16 (1)
1.3550E+07	1.2963E-13	44201986.6	*8.8990E-16*	3.59608E-14 (40.41)	8.89903E-16 (1)

Table 6.11.5-2: Planck's Energy (E^P = h.f)
Note: the above (error) is a ratio and therefore a value of 1 represents zero error

6.11.6 e, e'

The electrical charge in an electron is 1.60217648753E-19 C and designated the symbol 'e'. This charge is invariable; it never varies in an electron. This is not quite the same for the proton, however;

The charge generated in a proton by its orbiting electron is a constant and may be defined as follows:

e' = e.ξ_v.$\sqrt{[\underline{T}/\underline{T}_n]}$ = 2.75902141376572E-16 C

The static charge in a lone proton is always the same as that in an electron (e), but contrary to an electron, when a proton attracts an orbiting electron, its own charge will increase to e'. This is facilitated by the proton's surplus non-polar magnetic capacity (mass). I.e.:

The electrical charge in a lone proton (and in an electron) is always; e

The electrical charge in a proton with an orbiting electron rises to; e'

e' is also the maximum electrical charge in all the electro-magnetic radiation emitted by the proton

6.11.7 R_∞, R_γ, a_o

What is commonly regarded as Bohr's radius is the orbital radius of an electron that possesses the kinetic energy equal to Johannes Rydberg's predictions for an electron at what was then called an electron's ground state. 'a_o' is not, however, the ground state orbital radius of an electron.

$a_o = 4\pi.\varepsilon_o.\hbar^2 / m_e.e^2 = (h / 2\pi.m_e.c)^2 / R_n$ {m}

If the ground state of an electron can reasonably be said to occur at the point it ceases to provide sufficient energy to hold on to a neutron, this will be when the electron is orbiting:
at a radius of 8.3564315638157900E-09 m
at a temperature of 210.193328535837 K

Rydberg generated the following formula for his first constant:
$R_\infty = m_e.e^4 / 8.\varepsilon_o^2.h^3.c = 10973726.9561356$ {/m}
which breaks down to:
$R_\infty = \sqrt{R_n} / 4\pi.a_o^{1.5} = 10973726.9561356$ {/m}
and is his wave number (for electro-magnetic energy).

All of which breaks down to; $R_\infty = 1 / a_o.\xi_v$ {/m}

Rydberg generated the following formula for universal energy constant for an electron:
$R_\gamma = R_\infty.h.c.(Z.n)^2$
which breaks down to:
½ . R_n/a_o . $m_e.c^2 = 2.17987197684933E-18$ {J}
and relates to a temperature of 33192.4000063507 K
However, this value is of no significance to a key natural event.

But Planck's minimum orbital radius in his own constant 'h' is; $[4\pi]^2.a_o$
(refer to Chapters 5.4 & 6.11.5)

And if Rydberg had modified his formula to reflect Planck's value:
R_γ = ½ . (R_n / $[4\pi]^2.a_o$) . $m_e.c^2 = 1.38042005551962E-20$ {J}
Rydberg would have revealed the electron's minimum Planck energy 40 years earlier. Refer to Chapter 3.5.4

It is not yet understood what relevance Rydberg's radius (a_o) has to the atomic structure, because there is no such thing as 'rest-mass' for electrons as they never come to rest.

6.11.8 ρ_u

The ultimate (limiting) density is the maximum possible density achievable in nature, and it applies to both atomic particles; the electron and the proton.

ρ_u = 7.12660796350449E+16 kg/m³

It may be calculated as follows:

$\rho_u = \sqrt{[m_e.m_p]} / V_e^P$
Refer to Chapter 6.10.2 for V_e^P

or

$\rho_u = a_o.c^2 / G$
Refer to Chapter 6.11.2

or

$\rho_u = \sqrt{[m_e.m_p / \Sigma]}$

6.11.9 R_s

Schwarzschild's radius is the radius of a spherical body, the non-polar magnetic (gravitational) energy of which, is sufficient at its surface to trap the mass of an electron travelling at light-speed.

It is calculated as follows:

$R_s = 2.G.m/c^2$
Refer to Chapter 5 for definitions of 'G' and 'c'
m is the mass of the spherical body

The following examples give some idea of the Schwarzschild radii of various objects:

$1m^3$ of matter of ultimate density (7.12660796350450E+16 kg):
$R_s = 2.G.m_p/c^2 = 1.05835442134E-10$ m
i.e. you don't need very much of this matter to trap an electron travelling at light-speed.

The minimum sized black-body of iron density (ρ_i):
$R_s = c.\sqrt{[\ 3\ /\ 8.\pi.G.\rho_i\ ]} = 1.42875013455622E+11$ m
$m = {}^4/_3\pi R_s^3$. $\rho_i = 9.6237854E+37$ kg

A proton:
$R_s = 2.G.m/c^2 = 2.48396784934951E-54$ m
Proving that a proton cannot trap an electron through gravitational force alone. But if the coupling ratio were applied, obtaining potential energy through electrical charge;
$R_s = 2.G.m\ /\ \varphi.c^2 = 5.63587590767792E-15$ m
and the neutronic radius is: $R_n = 2.81793795383896E-15$ (exactly half!)

Given that the Schwarzschild's radius is sized to trap kinetic energy, and potential energy is exactly twice kinetic energy in orbiting electrons, half Schwarzschild's radius is the limiting potential electrical [charge] energy required to trap an electron travelling at light-speed.

A Schwarzschild's radius is, however, an hypothetical value as an electron travelling at velocity 'c' is supposed to represent a photon, which doesn't exist, so trapping a '*photon*' is an unreal scenario.

6.11.10 R_n

The neutronic radius is the orbital radius of an electron when it is travelling at the speed of electro-magnetic radiation (light-speed). At this speed, the electron and proton combine to become a neutron.

$R_n = \mu'.e.RC$ { kg.m / C² . C . C/kg = m }

According to Newton's orbital motion formula; $v = \sqrt{[R.g]}$
'v' will be the speed of light (c) (Table 10)
when:
R_n = orbital radius of $1.46677550700177 \times (r_p + r_e)$
    = 2.817937953839E-15 m
$g = G.m_p / \varphi.R_n^2 = 3.18940728807829E+31$ m/s²

Moreover, if:
$T = X.v^2 = X_R/R$
$R = X_R / X.v^2$
At light-speed (c)
$R_n = X_R / X.c^2 = 2.81793795383896E-15$ m

According to the relationship: $\underline{T} = X.v^2$
at a temperature of $\underline{T} = X.c^2 = 623316124.71718$ K
The reason an electron is trapped at this radius, is due to it venturing inside the Schwarzschild radius of the proton at this velocity (refer to Chapter 6.11.9).

6.11.11 RAC & RAM

Relative atomic charge and relative atomic mass define the capacity of a particle to hold an electrical charge (e) or *mass* (m) respectively.
They are related as follows:

RAM {g/mol}

RAC {C/mol}

If either of the above is divided by Avogadro's constant (N_A), you will get the capacity (m, e) of the particle {g, C}

If ideal gas constant (R_i) is divided by either of the above, you will get the *specific* capacity (c, q) of the particle {J/g/K, J/C/K}

If the capacity is multiplied by the *specific* capacity, you will get the *relative* capacity (C, Q) of the particle {J/K}

6.11.12 N_A

Avogadro's number is the number of C^{12} (pure carbon) atoms in 12g and is recognised as; N_A = 6.02214129E+23 /mole

However, one atom of pure carbon-12 has a mass of:
m_C = 6.($m_e+m_p+m_n$) = 2.00823909216000E-23 g

i.e.; N_A = 1/(m_e+m_p) {/mol}
where m_e & m_p are specified in grams

and 12 grams of pure carbon-12 contains:
N_A = 12/m_C = 5.97538412973187E+23
which is 0.7764208% less than Avogadro's number

If corrected, this would, of course alter a number of constants such as:
R_i; X; X_R; c & q

However, because this 0.78% inconsistency has little effect on the Newton's laws of motion, the corrected figure has not been adopted here. Avogadro's Number has been left as he defined it.

6.11.13 k_B

Boltzmann's constant describes the potential energy per unit temperature between a proton and its orbiting electron, and comprises the following fundamental constants:

$k_B = m_e.c^2 / Y.T_n = 1.38065156\text{E-}23$ J/K

Together with Avogadro's constant, Boltzmann's constant defines the ideal gas constant:

$R_i = N_A.k_B = 8.24992342031355$ J/K/mol

Note:
$e^2/4\pi$ is what is commonly referred to as the *'fine structure constant'* (α)

6.11.14   K

This constant of proportionality is common to all orbits.

In elliptical orbits, its value is governed by the *force-center's* mass and its formula is not only;

$$K = t^2/a^3 = (2.\pi)^2 / G.m_1$$

but also;

$$K = 2\pi/v_{max} \cdot 2\pi/v_{min} / a \; \{s^2/m^3\}$$

Where:
v_{max} = maximum orbital velocity
v_{min} = minimum orbital velocity
a = half the orbital major axis

In circular orbits, its value is governed by the *satellite's* mass, and its formula alters to:

$$K = (2\pi/v)^2 / R \; \{s^2/m^3\}$$
Where:
v = orbital velocity
R = orbital radius

Its value can be any number greater than 0

For a proton: K = 0.15587874533403

For our lunar system: K = 9.91826542816423E-14

For our solar system: K = 2.97491436434708E-19

For our Milky Way: K = 3.35025744566253E-30
Assuming an orbital eccentricity of 0.015941744 for our solar system

6.11.15 v_C & v_E

Two velocity constants (v_C & v_E) have been identified, but neither have yet been fully resolved in terms of their orbital applicability.

If $E = 3.k_B/m_n$:
$v_C = \sqrt{[\ k_B.3/m_n\ ]} = 157.320597065167$ {m/s}
or
$v_C = \sqrt{[\ g.R/PE\ .\ (E_1.E_2)/(E_1+E_2)\ ]} = \mathbf{157.320361123578}$ {m/s}
where:
$E_1 = E.m_{sun}$
$E_2 = E.m_{earth}$;
m_n = mass of a neutron
m_{sun} = mass of our sun
m_{earth} = mass of the earth

Using the values for a proton-electron pair and its neutron as its mass is most accurately known:
$v_E = \sqrt{[\ k_B'.T/m_e\ ]}$
@ 1K: $v_E = \mathbf{12007.8850946582}$ {m/s}
Where: k_B' is the charge level of a proton (refer to Chapters 5.4 & 6.11.13)

The velocity of any electron in any shell is found by applying the correct temperature value (T) in the above formula.

Whilst both of the above velocities are genuine constants that can be applied to all orbits (circular and elliptical), their resolution will feature in a later revision of this book.

6.12 The Laws of Thermodynamics

The First Law of Thermodynamics: Conservation of energy
Energy can never be lost, it can only be transformed or transferred.

The Second Law of Thermodynamics: Heat will not spontaneously pass from a colder body to hotter body

A high-energy source (hotter body) will spontaneously lose energy to a low-energy source (colder body) but you must add work if you want energy to transfer in the other direction (up-hill so to speak). This law essentially states that it is impossible to create energy from nothing.

This law also claims that energy can, and in fact is, lost by a system to its surroundings but that the reverse cannot happen i.e. an increase in disorder is an inevitable feature of time.

The Third Law of Thermodynamics: The entropy of a substance approaches zero as its temperature approaches zero (absolute)

Entropy is the term used to define disorder. The higher a substance's temperature the more disordered will be its atomic structure and the higher its entropy. E.g. gas has a higher entropy than a solid substance.

7 Things You Can Do

After two and a half years intensive study of this subject I'm knackered! I have left the last bits for somebody else to solve: you perhaps!

[1] Neutrons

Refer to: *The Life & Times of a Neutron*; Keith Dixon-Roche; published by CalQlata

[2] Nucleic Structure

The nucleic structure (ζ : refer to Chapter 3.5.3) is mirrored by the lattice structure of an atom in both viscous and its gaseous forms. The mathematical association of ζ & Γ with the lattice are yet to be resolved.

[3] T_o

The minimum Planck temperature may be related to neutron retention. This needs to be confirmed by experiment.

[4] T_c

The temperature at which a proton is anticipated to lose its orbiting electron. This needs to be confirmed by experiment.

[5] Electrons in Free-Flight

Whilst electrons cannot *lose* kinetic energy whilst in free-flight, it is not known whether they can collect energy from surrounding radiation. This needs to be confirmed by experiment.

[6] Electro-Magnetic Radiation

It is currently *assumed* that an electro-magnetic wave can donate *part* of its energy thereby altering its characteristics. This needs to be confirmed.

[7] Negative Velocity

It is possible, but by no means probable, that the negative-positive balance between all properties means that a negative velocity must also exist. This would of course mean that it could be possible to travel backwards; i.e. reverse time! But it needs to be proven.

Appendices

References, symbols, glossary, etc. used throughout this book along with a summary list of corollaries and hypotheses.

PHILOSOPHIÆ NATURALIS PRINCIPIA MATHEMATICA Revision IV

A1 General

N/A

A2 References

Most of the references used for the creation of this book are from original work supplied in CalQlata (www.calqlata.com), but some additional sources are listed below:

Magnificent Principia; Colin Pask; 978-1-61614-745-7

Seven Brief Lessons on Physics; Carlo Rovelli; 978-0-141-98172-7

Science Data Book; Open University; 0 05 002487 6

Science and Technology Dictionary; Chambers; 0-550-18026-5

A Dictionary of Scientific Units; H G Jerrard & D B McNeill; 0-412-28100-7

It is important to note here that most of the sources here are from work done by pre-20th Century scientists that are universally known and available from sources too numerous to mention here.

A3 Glossary

Atomic Particle	One of the three components that comprise an atom
Big-Bang	The eruption that occurred when the Ultimate-Body accumulated sufficient 'mass' to compromise the integrity of the innermost neutron.
Black-Body	A collection of Quanta too cold to emit electro-magnetic radiation in the frequencies that would enable detection.
Coupling Ratio (φ)	The ratio of the coupling force due to a magnetic charge and the coupling force due to an electric charge: $\varphi = G.m_p.m_e \div k.e^2$ (refer to Chapter 6.11.3)
Gas	Atoms that possess greater electrical field energy than magnetic field energy
Hades	The Milky Way's force-centre
Proton-electron pair	A proton that hosts an orbiting electron
Sub-Atomic Particle	The many particles said to compromise atomic particles (leptons, gluons, fermions, quarks, etc.)
Ultimate-Body	A body that contains all the Quanta in the universe ($\approx$2.8E+75) and represents the maximum single 'mass' that can exist without generating a Big-Bang.
Ultimate Density	The mass-density of all three atomic particles $\rho = 7.12660796350449E+16$ kg/m^3 Nothing in nature has a 'mass'-density greater than this value
Universal Period	The time elapsed since the last Big-Bang or between subsequent Big-Bangs
Viscous	Solid or liquid matter in which magnetic field energy is greater than an electron's electrical charge

All other definitions can be found on the following web page:
http://calqlata.com/help_definitions.html

A4 Symbols

Refer to Chapter 5 for a list of all the symbols used in this book.

The most prominent subscripts are listed below:

mass	e	electron
	p	proton
temperature	c	cold
	o	minimum Planck
	m	mean Planck
	n	maximum Planck
Rydberg	γ	energy constant
	∞	wave number
	o	orbital radius (a_o) *occasionally referred to as the Bohr radius*
Others	u	Ultimate
radii	n	Neutron orbit radius
	s	Schwarzschild radius
	1	force-centre
	2	satellite
energy	e	electron
	p	proton

The most prominent superscripts are listed below:

Force	N	Newton
	P	Planck

A5 Useful Formulas

Equidistant arc-length between 'n' points on the surface of a sphere:
$d = \pi.A / C.n$
where C is the circumference of the sphere
Linear distance across arc-length 'd' (above):
$\ell = 2.R.\mathrm{Sin}(\tfrac{1}{2}.d/R)$
but if you know 'ℓ' and need to find 'n':
$n = \pi / \mathrm{Asin}(\tfrac{1}{2}.\ell/R)$
and if $\ell=R$:
$n = \pi / \mathrm{Asin}(\tfrac{1}{2}) = \mathbf{6}$

Lorentz's Equation (magnetic force or field strength):
$F = q.v.B$
Which becomes:
$F = q.g.R.B$
for the laws of orbital motion
Where:
q is the total electrical charge $= q_1.q_2 / m_e.(q_1+q_2)$
v = relative velocity (electrical circuits)
g = gravitational attraction between m_1 & m_2
R = radial separation between m_1 & m_2
$B = \mu_o.I / 2.\pi.R$ kg/C {$R = 2.R_n$}
$I = e$
$B = \mu_o.e / 4.\pi.R_n = \cancel{4.\pi.R_n}.m_e/e^2 \cdot e / \cancel{4.\pi.R_n} = m_e/e = 1/RC$ kg/C
RC_e is the relative atomic charge of an electron {C/kg}
$B = 1/RC = 5.685634367312\mathrm{E}{-}12$ kg/C

PHILOSOPHIÆ NATURALIS PRINCIPIA MATHEMATICA Revision IV

A6 Corollaries

Corollary 1: Everything in the universe is composed of electrical and magnetic energy

Corollary 2: Magnetic energy is accrued and travels from positive to negative

Corollary 3: Electrical energy is shared and travels from negative to positive

Corollary 4: Atoms comprise collections of proton-electron pairs

Corollary 5: A neutron is a proton-electron pair united under high temperature and holds 4.0935556113127E-14 J of energy

Corollary 6: Atoms exist in solid/liquid state (attraction) due to the magnetic field generated by its proton-electron pairs

Corollary 7: Atoms exist in gaseous state (repulsion) due to the electrical field generated by its proton-electron pairs

Corollary 8: All matter is either viscous (solid/liquid) or gaseous, dependent upon the dominance of its magnetic or electrical fields

Corollary 9: Mass is non-polar magnetic charge

Corollary 10: Gravity is the attractive force between magnetic charges

Corollary 11: Light is electro-magnetic energy

Corollary 12: Every orbital system *must* have a force-centre

Corollary 13: Isaac Newton's gravitational constant may be defined as follows:
$G = a_o.c^2/\rho_u$ {m³ / s².kg per m³}
Where: ρ_u is the ultimate density (7.12661E+16 kg/m³)

Corollary 14: Potential energy remains constant irrespective of distance from its source

Corollary 15: Orbital shape is defined *only* by force-centre mass

Corollary 16: Kinetic energy of a satellite in Newton's laws of orbital motion is exclusive of that generated by its angular velocity

PHILOSOPHIÆ NATURALIS PRINCIPIA MATHEMATICA Revision IV

Corollary 17a: $PE/KE = -2.(1-e)/(1-e^2)$ for all orbits
Corollary 17b: The potential energy between a satellite and its force-centre is always twice the kinetic energy in the satellite for circular orbits

Corollary 18: The internal pressure of any mass can be calculated using Isaac Newton's force-formula thus:
$p = G.m_1.m_2 / A^2$, in which 'A' is the spherical area at radius 'r' from its centre (G must be in its modified form: 8.3862834423E-10 m³ / s².kg).

Corollary 19: The centrifugal force on an orbiting body:
@ the perigee of an ellipse; $F_c = F / (1+e)$
@ the apogee of an ellipse; $F_c = F . (1+e)$

Corollary 20: The centrifugal velocity (v_c) of a satellite at any point in an elliptical orbit is:
$v_c = ζ.v$
Where:
$ζ = \sqrt{[(f.Sin(θ/2)^α + p.Cos(θ/2)^α) / (f.cos(θ/2)^α + p.Sin(θ/2)^α)]}$
$α = \sqrt{[^4/_3.π]}$
v = the satellite's elliptical velocity at the same point.

Corollary 21: A satellite's spin is defined by its force-centre's spin and its sub-satellite orbits

Corollary 22a#: Prograde spin is induced in a satellite by the potential energy between its centre and that of its force-centre
Corollary 22b#: Retrograde spin is induced in a satellite by prograde spin energy in its force-centre
Corollary 22c#: Prograde spin is induced in a force-centre by the sum of the perigee kinetic energies plus apogee potential energies of its satellite(s)
Assuming a satellite's orbit is in the prograde direction#:

Corollary 23: The difference between the spin rates in a satellite's core and its mantle is due to the conflicting influences of corollary 22

Corollary 24a: A satellite's internal heat is generated by the friction due to corollary 23
Corollary 24b: Heat can be generated within a satellite orbiting a force-centre with a period different to the force-centre's period of spin
Corollary 24c: Heat is increased within a satellite that has satellites of its own
Corollary 24d: A gas-planet is a satellite with sufficient internal heat to prevent a surface crust forming

Corollary 25: A satellite's mantle plumes are generated by its internal frictional heat.

Corollary 26: A satellite's magnetic field is generated by the differential spin rates in its core and its mantle.

Corollary 27: The angular difference between true and magnetic axes in a satellite with iron core and mantle that rotate at different rates can be calculated thus:
$\alpha = \text{sign}(|\omega/\omega_m|) \cdot \tfrac{1}{2} \cdot \sqrt{[\text{Asin}(|\omega/\omega_m|)]}$

Corollary 28: All stars produce hydrogen gas in the form of proton-electron pairs, by converting proton-electron pairs (in its body-matter) to neutrons using the frictional heat generated by planetary spin

Corollary 29: Any satellite may be replaced in any orbit without altering the orbital shape and period (velocities)

Corollary 30: A black-body is a celestial body of any mass or size that is too cold to emit electro-magnetic radiation

Corollary 31: The force-centre at the heart of the Milky Way galaxy (Hades) has a mass of $\approx 1.8E+41$ kg, a diameter of $3.5E+12$ m and is spinning at $\approx 2E-07$ c/s according to NASA data on the Milky Way

Corollary 32: The density of the matter beneath the earth's crust is little more than that of water

Corollary 33: The coupling ratio (φ), that of magnetic to electrical potential energy is; $4.407E-40$

Corollary 34: There is insufficient pressure at the core of a minimum celestial black-body to alter the density of matter.

Corollary 35: Satellites in circular orbits generate their own kinetic energy.

Corollary 36a: All heat is radiated
Corollary 36b: Conduction is the transfer of electro-magnetic energy between electrons within matter irrespective of its state
Corollary 35c: Convection is the movement of gaseous atoms to balance electrical repulsive forces (between adjacent atoms) with gravitational forces

Corollary 37: E=mc² applies to potential energy in circular orbits

Corollary 38: Mass remains constant irrespective of velocity

Corollary 39: There is no such thing as a photon
Electrons in free-flight do not emit light – light possesses no mass

Corollary 40: There is no such thing as dark matter in the form of sub-atomic particles

A7 Hypotheses

Hypothesis 1: There is no such thing as mass.

Hypothesis 2: The earth's magnetic field reverses when a galactic comet passes sufficiently close to tip the earth on its axis.

Hypothesis 3: Electrons gain kinetic energy from electro-magnetic radiation, but they can only lose it via proton-electron pairing or impact.

Hypothesis 4: A lattice structure is mirrored in the atomic nucleic matrix.

Hypothesis 5: Only atoms of identical nucleic construction can generate lattice structures.

Hypothesis 6: An element's lattice structure also applies to its gaseous form and is responsible for partial pressure.

A8 The Heroes

The heroes of this story, to which I offer my gratitude, are listed below

It is not necessary to identify the invaluable contributions made by each of these contributors, they are all widely known and available in almost every scientific publication in circulation today.

Nicolaus Copernicus (Polish) 1473-1543
William Gilbert (English) 1544-1603
Tyco Brahe (Danish) 1546-1601
Galileo Galilei (Italian) 1564-1642
Johannes Kepler (German) 1571-1630
Christiaan Huygens (Dutch) 1629-1695
Isaac Newton (English) 1642-1727
Edmund Halley (English) 1656-1741
Charles-Augustin de Coulomb (French) 1736-1806
Hans Christian Ørsted (Danish) 1777-1851
Michael Faraday (English) 1791-1867
Josef Stefan (Austria) 1815-1863
James Clerk Maxwell (Scottish) 1831-1879
William Crookes (English) 1832-1919
Ludwig Boltzmann (Austria) 1844-1906
Hendrik Lorentz (Dutch) 1853-1928
Jules Henri Poincaré (French) 1854-1912
Johannes Robert Rydberg (Swedish) 1854-1919
Max Karl Ernst Ludwig Planck (German) 1858-1947

The others that were instrumental in the completion of this book are:

My long-suffering wife (Brigitte) sub-editor and critic

My daughter (Eléonore), who initiated this project

Kenneth Pickering friend & editor, who first suggested that I write it

My thanks go out to all the above each of whom have provided a valuable piece of the puzzle without which the final solution would not have been possible, along with my sincere apologies to anybody I have unintentionally omitted.